James .... he *New York Times*. .... has been translated into thirty languages. He lives in New York.

Praise for *Isaac Newton*:

'Brilliantly orchestrated and compelling ... a masterpiece'

MICHAEL HOLROYD

'It's beautifully paced and very stylishly written: compact, atmospheric, elegant. It offers a brilliant and engaging study in the paradoxes of the scientific imagination.' RICHARD HOLMES

'[Gleick's] *Isaac Newton* will, I think, set a new benchmark for our understanding of Newton the man ... Gleick is a prize-winning popular-science writer, with a keen understanding of the place of science in the modern world, and the ability to make its technicalities vivid and immediate for a lay audience ... [He] has given us a refreshingly bold version of Newton, a captivating model for mankind's struggle with the secrets of nature.'

LISA JARDINE, *Literary Review*

'Wonderful ... leaves you hungry for more' *Independent*

'[Gleick's] admirable new biography is perhaps the most accessible to date. He is an elegant writer, brisk without being shallow, excellent on the essence of the work, and revealing in his account of Newton's dealings with the times and with the men ... with whom he condescended to interact.' *Financial Times* magazine

BY THE SAME AUTHOR

Chaos: Making a New Science
Genius: The Life and Science of Richard Feynman
Faster: The Acceleration of Just About Everything
What Just Happened: A Chronicle from the Information Frontier

JAMES GLEICK

Isaac Newton

HARPER PERENNIAL

Harper Perennial
An imprint of HarperCollins*Publishers*
77-85 Fulham Palace Road
Hammersmith
London w6 8jb

www.harpercollins.co.uk/harperperennial

This edition published by Harper Perennial 2004
3

First published in Great Britain by Fourth Estate 2003

A catalogue record for this book is available from the British Library

ISBN 978-0-00-716318-2

Set in Monotype Janson by
Rowland Phototypesetting Ltd
Bury St Edmunds, Suffolk

Printed and bound in Great Britain by
Clays Ltd, St Ives plc

To Toby, Caleb, Asher and Will

I asked him where he had it made, he said he made it himself, & when I asked him where he got his tools said he made them himself & laughing added if I had staid for other people to make my tools & things for me, I had never made anything . . .

Contents

Illustrations

*By permission of the syndics of Cambridge University.

Isaac Newton

Isaac Newton said he had seen further by standing on the shoulders of giants, but he did not believe it. He was born into a world of darkness, obscurity and magic; led a strangely pure and obsessive life, lacking parents, lovers and friends; quarrelled bitterly with great men who crossed his path; veered at least once to the brink of madness; cloaked his work in secrecy; and yet discovered more of the essential core of human knowledge than anyone before or after. He was chief architect of the modern world. He answered the ancient philosophical riddles of light and motion, and he effectively discovered gravity. He showed how to predict the courses of heavenly bodies and so established our place in the cosmos. He made knowledge a thing of substance: quantitative and exact. He established principles, and they are called his laws.

Solitude was the essential part of his genius. As a youth he assimilated or rediscovered most of the mathematics known to humankind and then invented the calculus – the machinery by which the modern world understands change and flow – but kept this treasure to himself. He embraced his isolation through his productive years, devoting himself to the most secret of sciences, alchemy. He feared the light of exposure, shrank from criticism and controversy, and seldom published his work at all. Striving to decipher the riddles of the universe, he emulated the complex secrecy in which he saw them encoded. He stood aloof from other philosophers even after becoming a national icon – Sir Isaac, Master of the Mint, President of the Royal Society, his likeness engraved on medals, his discoveries exalted in verse.

'I don't know what I may seem to the world,' he said before he died, 'but, as to myself, I seem to have been only like a boy playing on the sea-shore, and diverting myself in now and then finding a smoother pebble or a prettier shell than ordinary, whilst the great ocean of truth lay all undiscovered before me.'' An evocative simile, much quoted in the centuries that followed, but Newton never played at the seashore, boy or man. Born in a remote country village, the son of an illiterate farmer, he lived in an island nation and explained how the moon and sun tug at the seas to create tides, but he probably never set eyes on the ocean. He understood the sea by abstraction and computation.

His life's path across the earth's surface covered barely 150 miles: from a hamlet of rural Lincolnshire southward to the university town of Cambridge and thence to London. He was born in the bedchamber of a stone farmhouse on Christmas 1642 (as the calendar was reckoned in England – but the calendar was drifting out of step with the sun). His father, Isaac Newton, yeoman, had married at thirty-five, fallen ill, and died before his son's birth. English had a word for that: the child was posthumous, thought unlikely to resemble the father.

This first Isaac Newton left little trace: some sheep, barley, and simple furniture. He endorsed his will with his X, for like most of his countrymen he could neither read nor write. He had worked the land of Woolsthorpe, a place of woods, open heaths, brooks and springs, where underneath the thin soil lay a grey limestone, from which a few dwellings were built to last longer than the common huts of timber and clay. A road of the Roman Empire passed nearby, running south and north, a reminder of ancient technology still unsurpassed. Sometimes children unearthed antique coins or remains of a villa or wall.[2]

The second Isaac Newton lived to be eighty-four, gouty and rich. He died in London at the end of the winter of 1727, a prolonged and excruciating death from a kidney stone. England for the first time granted a state funeral to a subject whose attainment lay in the realm of the mind. The Lord Chancellor, two dukes and three earls bore the pall, with most of the Royal Society following behind. The corpse lay in state in Westminster Abbey for eight days and was buried in its nave. Above the grave was carved an ornate monument in grey and white marble: the figure of Newton, recumbent; the celestial globe, marked with the path of a 1680 comet; and angelic boys playing with a prism and weighing the sun and planets. A Latin inscription hailed his 'strength of mind almost divine' and 'mathematical principles peculiarly his own' and declared: 'Mortals rejoice that there has existed so great an ornament of the human race.' For England, the continent of Europe, and then the rest of the world, Newton's story was beginning.

The French writer calling himself Voltaire had just reached London. He was amazed by the kingly funeral and exhilarated by all things Newtonian. 'A Frenchman arriving in London finds things very different,' he reported. 'For us it is the pressure of the moon that causes the tides of the sea; for the English it is the sea that gravitates towards the moon, so that when you think that the moon should give us a high tide, these gentlemen think you should have a low one.' It pleased Voltaire to compare Newton with his nation's late philosophical hero, René Descartes: 'For your Cartesians everything is moved by an impulsion you don't really understand, for Mr Newton it is by gravitation, the cause of which is hardly better known.' The most fundamental conceptions were new and up for grabs in coffee houses and salons. 'In Paris you see the earth shaped like a melon, in London it is flattened on two sides. For a Cartesian light exists in the air, for a Newtonian it comes from the sun in six and a half minutes.' Descartes was a dreamer; Newton a sage. Descartes experienced poetry and love; Newton did not. 'In the course of such a long life he had neither passion nor weakness; he never went near any woman. I have had that confirmed by the doctor and the surgeon who were with him when he died.'³

What Newton learned remains the essence of what we know, as if by our own intuition. Newton's laws are our laws. We are Newtonians, fervent and devout, when we speak of forces and masses, of action and reaction; when we say that a sports team or political candidate has momentum; when we note the inertia of a tradition or bureaucracy; and when we stretch out an arm and feel the force of gravity all around, pulling earthward. Pre-Newtonians did not feel such a force. Before Newton the English word *gravity* denoted a mood –

seriousness, solemnity – or an intrinsic quality. Objects could have heaviness or lightness, and the heavy ones tended downward, where they belonged.[4]

We have assimilated Newtonianism as knowledge and as faith. We believe our scientists when they compute the past and future tracks of comets and spaceships. What is more, we know they do this not by magic but by mere technique. 'The landscape has been so totally changed, the ways of thinking have been so deeply affected, that it is very hard to get hold of what it was like before,' said the cosmologist and relativist Hermann Bondi. 'It is very hard to realise how total a change in outlook he produced.'[5] Creation, Newton saw, unfolds from simple rules, patterns iterated over unlimited distances. So we seek mathematical laws for economic cycles and human behaviour. We deem the universe solvable.

He began with foundation stones of knowledge: time, space, motion. *I do not define time, space, place, and motion, as being well known to all*, he wrote in mid-life – then a reclusive professor, recondite theologian and alchemist, seldom leaving his room in Trinity College, Cambridge.[6] But he did mean to define these terms. He salvaged them from the haze of everyday language. He standardised them. In defining them, he married them, each to the others.

He dipped his quill in an ink of oak galls and wrote a minuscule Latin script, crowding the words edge to edge: *The common people conceive those quantities under no other notions but from the relation they bear to sensible objects. And thence arise certain prejudices . . .* By then he had written more than a million words and published almost none. He wrote for himself, careless of food and sleep. He wrote to calculate, laying down numbers in spidery lines and broad columns. He computed as most people daydream. The flow of his thought slipped back and forth between English and Latin. He wrote to read, copying out books and manuscripts verbatim, sometimes the same text again and again. More determined than joyful, he wrote to reason, to meditate, and to occupy his febrile mind.

His name betokens a system of the world. But for Newton himself there was no completeness, only a questing – dynamic, protean, and unfinished. He never fully detached matter and space from God. He never purged occult, hidden, mystical qualities from his vision of nature. He sought order and believed in order but never averted his eyes from the chaos. He of all people was no Newtonian.

Information flowed faintly and perishably then, through the still small human species, but he created a method and a language that triumphed in his lifetime and gained ascendancy with each passing century. He pushed open a door that led to a new universe: set in absolute time and space, at once measureless and measurable, furnished with science and machines, ruled by industry and natural law. Geometry and motion, motion and geometry: Newton joined them as one. With the coming of Einstein's relativity, Newtonian science was often said to have been 'overthrown' or 'replaced', but that was not so. It had been buttressed and extended.[7]

'Fortunate Newton, happy childhood of science!' said Einstein. 'Nature to him was an open book. He stands before us strong, certain, and alone.'[8]

Yet he speaks to us reluctantly and covertly.

I

WHAT IMPLOYMENT
IS HE FIT FOR?

Medieval, in some disrepair, the Woolsthorpe farmhouse nestled into a hill near the River Witham. With its short front door and shuttered windows, its working kitchen, and its bare floors of ash and linden laid on reeds, it had belonged to Newton's forebears for just twenty years. At the back stood apple trees. Sheep grazed for acres around.

Isaac was born in a small room at the top of the stairs. By the terms of feudal law this house was a manor and the fatherless boy was its lord, with seigniorial authority over a handful of tenant farmers in nearby cottages. He could not trace his ancestry back past his grandfather, Robert, who lay buried in the churchyard nearly a mile to the east. Still, the boy expected to live managing the farm in the place of the father he had never known. His mother, Hannah Ayscough, had come from gentlefolk. Her brother, the Reverend William Ayscough, studied at Cambridge University on his way to joining the Anglican clergy; now he occupied a village rectory two miles away. When Isaac was three years old and his widowed mother near thirty, she accepted a marriage offer from another nearby rector, Barnabas Smith, a wealthy man twice her age. Smith wanted a wife, not a stepson; under the negotiated terms of their marriage Hannah abandoned Isaac in the Woolsthorpe house, leaving him to his grandmother's care.[1]

War flared in the countryside all through his youth. The decade-long Great Rebellion began in the year of his birth: Parliamentarians fighting Royalists, Puritans recoiling from the idolatry they saw in the Church of England. Motley, mercenary armies skirmished throughout the Midlands. Pikemen and musketeers sometimes passed through the fields near Woolsthorpe.[2] Bands of men plundered farms for supplies. England was at war with itself and also, increasingly, aware of itself – its nationhood, its specialness. Divided as it was, convulsed over ecclesiastical forms and beliefs, the nation carried out a true revolution. The triumphant Puritans rejected absolutism and denied the divine right of the monarchy. In 1649, soon after Isaac turned six, Charles Stuart, the king, was beheaded at the wall of his palace.

This rustic country covered a thousandth of the world's landmass, cut off from the main continent since the warming of the planet and the melting of polar ice 13,000 years before. Plundering, waterborne tribes had settled on its coasts in waves and diffused into its downs and valleys, where they aggregated in villages. What they knew or believed about nature depended in part on the uses of technology. They had learned to employ the power of water and wind to crush, grind and polish. The furnace, the forge and the mill had taken their place in an economy that thereby grew more specialised and hierarchical. People in England, as in many human communities, made metal – kettles of copper and brass, rods and nails of iron. They made glass. These crafts and materials were prerequisites now to a great leap in knowledge. Other prerequisites were lenses, paper and ink, mechanical clocks, numeric systems capable of denoting indefinitely small fractions, and postal services spanning hundreds of miles.

By the time of Newton's birth, one great city had formed, with about 400,000 people; no other town was even a tenth as large. England was still a country of villages and farms, its seasons ordered by the Christian calendar and the rhythms of agriculture: lambing and calving, haymaking and harvest. Years of harvest failure brought widespread starvation.[3] Roving labourers and vagrants made up much of the population. But a class of artisans and merchants was coming into its own: traders, shopkeepers, apothecaries, glaziers, carpenters and surveyors, all developing a practical, mechanical view of knowledge.[4] They used numbers and made tools. The nucleus of a manufacturing economy was taking shape.

When Isaac was old enough, he walked to the village dame school, where he learned to read and studied the Bible and chanted arithmetic tables. He was small for his age, lonely and abandoned. Sometimes he wished his stepfather dead, and his mother, too: in a rage he threatened to burn their house down over them. Sometimes he wished himself dead and knew the wish for a sin.[5]

On bright days sunlight crept along the wall. Darkness as well as light seemed to fall from the window – or was it from the eye? No one knew. The sun projected slant edges, a dynamic echo of the window frame in light and shadow, sometimes sharp and sometimes blurred, expressing a three-dimensional geometry of intersecting planes. The particulars were hard to visualise, though the sun was the most regular of heavenly objects, the one whose cycles already defined the measures of time. Isaac scratched crude geometric figures, circles with arcs inscribed, and hammered wooden pegs into the walls and the ground to measure time exactly, to the nearest quarter-hour.[6] He cut sundials into stone and charted the shadows cast by their gnomons. This meant seeing time as akin to space, duration as length, the length of an arc. He measured small distances with strings and made a translation between inches and minutes of an hour. He had to revise this translation methodically as the seasons changed. Across the day the sun rose and fell; across the year its position in the sky shifted slightly against the fixed stars and traced a slowly twisting figure eight,[7] a figure invisible except to the mind's eye. Isaac grew conscious of this pattern long before he understood it as the product of two oddities, the earth's elliptical orbit and a tilt in its axis.

At Woolsthorpe anyone who cared to know the hour consulted Isaac's dials.[8] 'O God! Methinks it were a happy life,' said Shakespeare's Henry VI, 'to carve out dials quaintly, point by point, thereby to see the minutes how they run.'[9] Sundials – shadow-clocks – still told most people the time, though at some churches the hour could be read from mechanical clocks. At night the stars turned in the blue vault of the sky; the moon waxed and waned and traced its own path, much like the sun's, yet not exactly – these great globes, ruling the seasons, lighting the day and night, connected as if by invisible cords.[10] Sundials embodied practical knowledge that had been refined over millennia. With cruder sundials, the hours were unequal and varied with the seasons. Better versions achieved precision and encouraged an altered sense of time itself: not just as a recurring cycle, or a mystical quality influencing events, but as duration, measurable, a dimension. Still, no one could perfect or even understand sundials until all the shifting pieces of a puzzle had been assembled: the shadows, the rhythms, the orbits of planets, the special geometry of the ellipse, the attraction of matter by matter. It was all one problem.

When Isaac was ten, in 1653, Barnabas Smith died, and Hannah returned to Woolsthorpe, bringing three new children with her. She sent Isaac off to school, eight miles up the Great North Road, to Grantham, a market town of a few hundred families – now a garrison town, too. Grantham had two inns, a church, a guild hall, an apothecary, and two mills for grinding corn and malt.[11] Eight miles was too far to walk each day; Isaac boarded with the apothecary, William Clarke, in the High Street. The boy slept in the garret and left signs of his presence, carving his name into the boards and drawing in charcoal on the walls: birds and beasts, men and ships, and pure abstract circles and triangles.[12]

At the Kings School – one room – with strict Puritan discipline, Henry Stokes, schoolmaster, taught eighty boys Latin, theology, and some Greek and Hebrew. In most English schools that would have been all, but Stokes added some practical arithmetic for his prospective farmers: mostly about measurement of areas and shapes, algorithms for surveying, marking fields by the chain, calculating acres (though the acre still varied from one county to the next, or according to the land's richness).[13] He offered a bit more than a farmer would need: how to inscribe regular polygons in a circle and compute the length of each side, as Archimedes had done to estimate pi. Isaac scratched Archimedes' diagrams in the wall. He entered the lowest form at the age of twelve, lonely, anxious and competitive. He fought with other boys in the churchyard; sometimes noses were bloodied. He filled a Latin exercise book with unselfconscious phrases, some copied, others invented, a grim stream of thought: *A little fellow; My poore help; Hee is paile; There is no room for me to sit; In the top of the house – In the bottom of hell; What imployment is he fit for? What is hee good for?*[14] He despaired. *I will make an end. I cannot but weepe. I know not what to doe.*

Barely sixty lifetimes had passed since people began to record knowledge as symbols on stone or parchment. One of England's earliest paper mills opened at the end of the sixteenth century, on the Deptford River. Paper was prized, and the written word played a small part in daily life. Most of what people thought remained unrecorded; most of what they recorded was hidden or lost. Yet to some it seemed a time of information surfeit. 'I hear new news every day,' wrote the vicar Robert Burton, attuned as he was – virtually living in the Bodleian Library at Oxford – to the transmission and storage of data:

those ordinary rumours of war, plagues, fires, inundations, thefts, murders, massacres, meteors, comets, spectrums, prodigies, apparitions, . . . and such like, which these tempestuous times afford . . . New books every day, pamphlets, currantoes, stories, whole catalogues of volumes of all sorts, new paradoxes, opinions, schisms, heresies, controversies in philosophy, religion &c.[15]

Burton was attempting to assemble all previous knowledge into a single rambling, discursive, encyclopedic book of his own. He made no apology for his resolute plagiarism; or, rather, he apologised this way: 'A dwarf standing on the shoulders of a Giant may see farther than a Giant himself.'[16] He tried to make sense of rare volumes from abroad, which proposed fantastic and contradictory schemes of the universe – from Tycho Brahe, Galileo, Kepler, and Copernicus. He tried to reconcile them with ancient wisdom.

Did the earth move? Copernicus had revived that notion, 'not as a truth, but a supposition'. Several others agreed. 'For if the Earth be the Center of the World, stand still, as most received opinion is',[17] and the celestial spheres revolve around it, then the heavens must move with implausible speed. This followed from measurements of the distance of sun and stars. Burton borrowed (and mangled) some arithmetic. 'A man could not ride so much ground, going 40 miles a day, in 2,904 years, as the Firmament goes in 24 hours; or so much in 203 years, as the said Firmament in one minute; which seems incredible.' People were looking at the stars through spyglasses; Burton himself had seen Jupiter through a glass eight feet long and agreed with Galileo that this wanderer had its own moons.

He was forced to consider issues of shifting viewpoint,

though there was no ready language for expressing such conundrums: 'If a man's eye were in the Firmament, he should not at all discern that great annual motion of the earth, but it would still appear an indivisible point.' If a man's eye could be so far away, why not a man? Imaginations ran free. 'If the earth move, it is a Planet, & shines to them in the Moon, & to the other Planetary Inhabitants, as the Moon and they to us upon the earth.'

We may likewise insert . . . there be infinite Worlds, and infinite earths or systems, in infinite æther, . . . and so, by consequence, there are infinite habitable worlds: what hinders? . . . It is a difficult knot to untie.

Especially difficult because so many different authorities threw forth so many hypotheses: our modern divines, those heathen philosophers, heretics, schismatics, the Church of Rome. 'Our latter Mathematicians have rolled all the stones that may be stirred: and . . . fabricated new systems of the World, out of their own Dædalean heads.'[18] Many races of men have studied the face of the sky throughout history, Burton said, and now the day was coming when God would reveal its hidden mysteries. Tempestuous times, indeed.

But *new books every day* did not find their way to rural Lincolnshire. Newton's stepfather, Smith, had owned books, on Christian subjects. The apothecary Clarke also owned books. Smith even possessed blank paper, in a large common-place book that he had kept for forty years. He painstakingly numbered the pages, inscribed theological headings on the first few, and otherwise left it almost entirely empty. Some time after his death this trove of paper came into Isaac's possession. Before that, in Grantham, with two and a half pence his mother had given him, Isaac was able to buy a tiny notebook, sewn sheets bound in vellum. He asserted his ownership with an inscription: *Isacus Newton hunc librum possidet.*[19] Over many months he filled the pages with meticulous script, the letters and numerals often less than one-sixteenth of an inch high. He began at both ends and worked towards the middle. Mainly he copied a book of secrets and magic printed in London several years earlier: John Bate's *Mysteryes of Nature and Art*, a scrap book, rambling and yet encyclopedic in its intent.

He copied instructions on drawing. 'Let the thing which you intend to draw stand before you, so the light be not hindered from falling upon it.' 'If you express the sunn make it riseing or setting behind some hill; but never express the moon or starrs but up on necessity.' He copied recipes for making colours and inks and salves and powders and waters. 'A sea colour. Take privet berries when the sun entreth into Libra, about the 13th of September, dry them in the sunn; then bruise them & steep them.' Colours fascinated him. He catalogued several dozen, finely and pragmatically distinguished: purple, crimson, green, another green, a light green, russet, a brown blue, 'colours for naked pictures', 'colours for dead corpes', charcoal black and seacoal black. He copied techniques for melting metal (in a shell), catching birds ('set black wine for them to drink where they come'), engraving on a flint, making pearls of chalk.

Living with Clarke, apothecary and chemist, he learned to grind with mortar and pestle; he practised roasting and boiling and mixing; he formed chemicals into pellets, to be dried in the sun. He wrote down cures, remedies, and admonitions:

THINGS HURTFULL FOR THE EYES

Garlick Onions & Leeks ... Gooing too suddaine after meals. Hot wines. Cold ayre ... Much blood-letting ... dust. ffire. much weeping ...

Bate's book mixed Aristotelianism and folklore: 'sundry Experiments both serviceable and delightfull, which because they are confusedly intermixed, I have entituled them *Extravagants*.' Isaac copied that word at the head of several pages. Bate described and illustrated many forms of water-works and fireworks, and Isaac spent hours cutting wood with his knife, building ingenious watermills and windmills. Grantham town was building a new mill; Isaac followed its progress and made a model, internalising the whirring and pounding of the machine and the principles that govern gears, levers, rollers and pulley wheels. In his garret he constructed a water clock, four feet high, from a wooden box, with an hour hand on a painted dial. He made paper lanterns. He crafted kites and sent them aloft at night trailing lanterns ablaze – lights in the black sky to frighten the neighbours.[20]

Bate offered knowledge as play, but with a nod to system: 'the four elements, Fier, Ayer, Water, and Earth, and the *prima Principia*,' he wrote. This venerable four-part scheme – with its corollary powers: dry, cool, warm, and moist – expressed a desire to organise, classify and name the world's elements, in the absence of mathematical and technological tools. Simple wisdom covered motion, too. Bate explained: 'Their light parts ascend upwards; and those that are more grosse & heavy, do the contrary.'[21]

Isaac omitted these principles from his copying. He crowded his tiny pages with astronomical tables related to sun-dialling, followed by an elaborate computation of the calendar for the next twenty-eight years. He copied lists of words, adding as many of his own as came to mind.[22] Across forty-two notebook pages he organized 2,400 nouns in columns under subject headings:

Artes, Trades, & Sciences: ... *Apothecary* ... *Armourer Astrologer Astronomer* ... *Diseases:* ... *Gobbertooth* ... *Gout* ... *Gangreene* ... *Gunshott* ... *Kindred, & Titles: Bridegroome* ... *Brother Bastard Barron* ... *Brawler Babler* ... *Brownist Benjamite* ... *Father Fornicator* ...

Thoughts of family were no balm to this troubled soul. Nevertheless, in the autumn of 1659, when Isaac was sixteen years old, his mother summoned him home to be a farmer.

2

SOME PHILOSOPHICAL QUESTIONS

He did not know what he wanted to be or do, but it was not tend sheep or follow the plough and the dung cart. He spent more time gathering herbs and lying with a book among the asphodel and moonwort, out of the household's sight.[1] He built water wheels in the stream while his sheep trampled the neighbours' barley. He watched the flow of water, over wood and around rocks, noting the whorls and eddies and waves, gaining a sense of fluid motion.[2] He defied his mother and scolded his half-sisters.[3] He was fined in the manor court for allowing his swine to trespass and his fences to lie in disrepair.[4]

His Grantham schoolmaster, Stokes, and his mother's brother, the rector William Ayscough, finally intervened. Ayscough had prepared for the clergy at the College of the Holy and Undivided Trinity, the greatest of the sixteen colleges at the University of Cambridge, so they arranged for Isaac to be sent there. He made the journey south, three days and two nights, and was admitted in June 1661. Cambridge recognised students in three categories: noblemen, who dined at high table, wore sophisticated gowns, and received degrees with little examination; pensioners, who paid for tuition and board and aimed, mainly, for the Anglican ministry; and sizars, who earned their keep by menial service to other students, running errands, waiting on them at meals, and eating their leftovers. The widowed Hannah Smith was wealthy now, by the standards of the countryside, but chose to provide her son little money; he entered Trinity College as a subsizar. He had enough for his immediate needs: a chamber pot; a notebook of 140 blank pages, three and a half by five and a half inches, with leather covers; 'a quart bottle and ink to fill it'; candles for many long nights, and a lock for his desk.[5] For a tutor he was assigned an indifferent scholar of Greek. Otherwise he kept to himself.

He felt learning as a form of obsession, a worthy pursuit, in God's service, but potentially prideful as well. He taught himself a shorthand of esoteric symbols – this served both to save paper and encrypt his writing – and he used it, at a moment of spiritual crisis, to record a catalogue of his sins. Among them were *neglecting to pray, negligence at the chapel*, and variations on the theme of falling short in piety and devotion. He rebuked himself for a dozen ways of breaching the Sabbath. On one Sunday he had whittled a quill pen and then lied about it. He confessed *uncleane thoughts words and actions and dreamese*. He regretted, or tried to regret, *setting my heart on money learning pleasure more than Thee*.[6] Money, learning, pleasure: three sirens calling his heart. Of these, neither money nor pleasure came in abundance.

The Civil War had ended and so had the Protectorate of Oliver Cromwell, dead from malaria, buried and then exhumed so his head could be stuck on a pole on top of

Westminster Hall. During the rebellion Puritan reformers had gained control of Cambridge and purged the colleges of many Royalist scholars. Now, with the restoration of Charles II to the crown, Puritans were purged, Cromwell was hanged in effigy, and the university's records from the Protectorate years were burned. This riverside town was a place of ferment, fifty miles from London, a hundredth its size, a crossroads for information and commerce. Each year between harvest and ploughing, tradesmen gathered for Stourbridge Fair, England's largest: a giant market for wool and hops, metalware and glassware, silk and stationery, books, toys and musical instruments – a bedlam of languages and apparel, and 'an Abstract of all sorts of mankind', as a pamphleteer described it.[7] Newton, scrupulous with his limited funds, bought books there and, one year, a glass prism – a toy, imprecisely ground, flawed with air bubbles. Often enough, the complex human traffic had another consequence: Cambridge suffered visitations of plague.

The curriculum had grown stagnant. It followed the scholastic tradition laid down in the university's medieval beginnings: the study of texts from disintegrated Mediterranean cultures, preserved in Christian and Islamic sanctuaries through a thousand years of European upheaval. The single authority in all the realms of secular knowledge was Aristotle – doctor's son, student of Plato, and collector of books. Logic, ethics and rhetoric were all his, and so – to the extent they were studied at all – were cosmology and mechanics. The Aristotelian canon enshrined systematisation and rigour, categories and rules. It formed an edifice of reason:

knowledge about knowledge. Supplemented by ancient poets and medieval divines, it was a complete education, which scarcely changed from generation to generation. Newton began by reading closely, but not finishing, the *Organon* and the *Nicomachean Ethics* ('For the things we have to learn before we can do them, we learn by doing them').[8]

He read Aristotle through a mist of changing languages, along with a body of commentary and disputation. The words crossed and overlapped. Aristotle's was a world of substances. A substance possesses qualities and properties, which taken together amount to a *form*, depending ultimately on its essence. Properties can change; we call this *motion*. Motion is action, change, and life. It is an indispensable partner of *time*; the one could not exist without the other. If we understood the cause of motion, we would understand the cause of the world.

For Aristotle motion included pushing, pulling, carrying and twirling; combining and separating; waxing and waning. Things in motion included a peach ripening, a fish swimming, water warming over a fire, a child growing into an adult, an apple falling from a tree.[9] The heavy thing and the light thing move to their proper positions: the light thing up and the heavy thing down.[10] Some motion is natural; some violent and un-natural. Both kinds revealed the connections between things. 'Everything that is in motion must be moved by something,' Aristotle asserted (and proved, by knotted logic).[11] A thing cannot be at once *mover* and *moved*. This simple truth implied a first mover, put in motion by no other, to break what must otherwise be an infinite loop:

Since everything that is in motion must be moved by something, let us take the case in which a thing is in locomotion and is moved by something that is itself in motion, . . . and that by something else, and so on continually: then the series cannot go on to infinity, but there must be some first mover.

To the Christian fathers, this first mover could only be God. It was a testament to how far pure reason could take a philosopher; and to how involuted and self-referential a chain of reasoning could become, with nothing to feed on but itself.

This all-embracing sense of motion left little place for quantity, measurement and number. If objects in motion could include a piece of bronze becoming a statue,[12] then philosophers were not ready to make fine distinctions, like the distinction between velocity and acceleration. Indeed, the Greeks had a principled resistance to mathematicising our corruptible, flawed, sublunary world. Geometry belonged to the celestial sphere; it might relate music and the stars, but projectiles of rock or metal were inappropriate objects for mathematical treatment. So technology, advancing, exposed Aristotelian mechanics as quaint and impotent. Gunners understood that a cannonball, once in flight, was no longer moved by anything but a ghostly memory of the explosion inside the iron barrel; and they were learning, roughly, to compute the trajectories of their projectiles. Pendulums, in clockwork, however crude, demanded a mathematical view of motion. And in turn the clockwork made measurement possible – first hours, then minutes. Of an object falling from a tower or rolling down an inclined plane, people could begin to ask: what is the distance? what is the time?

What, therefore, is the velocity? And how does the velocity, itself, change?

Nor was Aristotle's cosmology faring well outside Cambridge's gates. It was harmonious and immutable: crystalline spheres round the earth, solid and invisible, carrying the celestial orbs within them. Ptolemy had perfected his universe and then, for hundreds of years, Christian astronomers embraced and extended it, reconciled it with biblical scripture, and added a heaven of heavens, deep and pure, perhaps infinite, the home of God and angels, beyond the sphere of fixed stars. But as stargazers made increasingly detailed notations, they catalogued planetary motions too irregular for concentric spheres. They saw freaks and impurities, such as comets glowing and vanishing. By the 1660s – *new news every day* – readers of esoterica knew well enough that the earth was a planet and that the planets orbited the sun. Newton's notes began to include measurements of the apparent magnitude of stars.

Although the library of Trinity College had more than three thousand books, students could enter only in the company of a fellow. Still, Newton found his way to new ideas and polemics: from the French philosopher René Descartes, and the Italian astronomer Galileo Galilei, who had died in the year of Newton's birth. Descartes proposed a geometrical and mechanical philosophy. He imagined a universe filled throughout with invisible substance, forming great vortices that sweep the planets and stars forward. Galileo, meanwhile, applied geometrical thinking to the problem of motion. Both men defied Aristotle explicitly – Galileo by claiming that all bodies are made of the same stuff, which is heavy, and therefore fall at the same rate.

Not the same *speed*, however. After long gestation, Galileo created a concept of uniform acceleration. He considered motion as a state rather than a process. Without ever using a word such as *inertia*, he nonetheless conceived that bodies have a tendency to remain in motion or to remain motionless. The next step demanded experiment and measure. He measured time with a water clock. He rolled balls down ramps and concluded, wrongly, that their speed varied in proportion to the distance they rolled. Later, trying to understand free fall, he reached the modern definition, correctly assimilating units of distance, units of speed, and units of time. Newton began to absorb this, at second or third hand; Galileo had written mostly in Italian, a language few in England could read.[13]

In Newton's second year, having filled the beginning and end of his notebook with Aristotle, he started a new section deep inside: *Questiones quædam philosophicæ* – some philosophical questions. He set authority aside. Later he came back to this page and inscribed an epigraph borrowed from Aristotle's justification for dissenting from his teacher. Aristotle had said, 'Plato is my friend, but truth my greater friend.' Newton inserted Aristotle's name in sequence: *Amicus·Plato amicus Aristoteles magis amica veritas.*[14] He made a new beginning. He set down his knowledge of the world, organised under elemental headings, expressed as questions, based sometimes on his reading, sometimes on speculation. It showed how little was known, altogether. The choice of topics – forty-five in all – suggested a foundation for a new natural philosophy.

Of the First Matter. Of Atoms. Could he know, by the force of logic, whether matter was continuous and infinitely divisible, or discontinuous and discrete? Were its ultimate parts mathematical points or actual atoms? Since a mathematical point lacks body or dimension – 'is but an imaginary entity' – it seemed implausible that even an infinite number of them could combine to form matter with real extension,[15] even if bits of vacuum ('interspersed inanities') separated the parts. The question of God's role, as creator, could be dangerous territory. 'Tis a contradiction to say the first matter depends on some other subject' – in parentheses he added, 'except God'; then, on second thought, he crossed that out – 'since that implies some former matter on which it must depend.' Reasoning led him, as it had led ancient Greeks, to atoms – not by observation or experiment, but by eliminating alternatives. Newton declared himself a corpuscularian and an atomist. 'The first matter must be attoms. And that Matter may be so small as to be indiscernible.' Very small, but finite, not zero. Indiscernible, but unbreakable and indivisible. This was an unsettled conception, because Newton also saw a world of smooth change, of curves, and of flow. What about the smallest parts of time and motion? Were these continuous or discrete?

Quantity. Place. 'Extension is related to places, as time to days yeares &c.'[16] He invoked God on another controversial question: Is space finite or infinite? Not the imaginary abstract

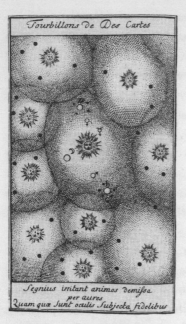

In the cosmos of Descartes, matter fills all space and forms whirling vortices.

space of geometers, but the real space in which we live. Infinite, surely! 'To say that extension is but indefinite' – Descartes said this, in fact – 'is as much to say God is but indefinitely perfect because wee cannot apprehend his whole perfection.'

Time and Eternity. No abstract disputation here; he just sketched a wheel-shaped clock, to be driven by water or sand, and raised wholly practical questions about making clocks with various materials, such as 'metalline globular dust'. Only then did he reach *Motion*, and again, he began by looking for the root constituents, the equivalent of atoms. *Motion* led to *Celestiall Matter & Orbes* – which took Newton, encountering the early echoes of Continental thought, to Descartes. In Descartes's universe, there could be no vacuum, for the universe was space, and space meant extension, and extension surely implied substance. Also, the world's principles were mechanical: all action propagated through contact, one object directly pushing another, no mystical influences from afar.

So a vacuum could not transmit light. Light was a form of

pression, Descartes said – imaginatively, because philosophers had barely begun to conceive of pressure as a quality that an invisible fluid, the air, could possess. But now Newton had heard of Robert Boyle's experiments with an air pump, and *pressure* was the word Boyle used in this new sense. Newton began again:

Whether Cartes his first element can turne about the vortex & yet drive the matter of it continually from the ☉ [sun] to produce light, & spend most of its motion in filling up the chinks between the globuli.[17]

From matter to motion, to light, and to the structure of the cosmos. The sun drove the vortex by its beams. The ubiquitous vortex could drive anything: Newton sketched some ideas for perpetual motion machines. But light itself played a delicate part in the Cartesian scheme, and Newton, attempting to take Descartes literally, already sensed contradictions. Pressure does not restrict itself to straight lines; vortices whirl around corners. 'Light cannot be by pression,' Newton asserted, 'for then wee should see in the night a[s] wel or better than in the day we should se[e] a bright light above us becaus we are pressed downewards . . .' Eclipses should never darken the sky. 'A man goeing or running would see in the night. When a fire or candle is extinguished we lookeing another way should see a light.'[18]

Another elusive word, *gravity*, began to appear in the *Questiones*. Its meanings darted here and there. It served as half of a linked pair: *Gravity & Levity*. It represented the tendency

Violent motion
(Newton's drawing).

of a body to descend, ever downwards. But how could this happen? 'The matter causing gravity must pass through all the pores of a body. It must ascend againe, for else the bowells of the earth must have had large cavitys & inanitys to containe it in . . .' [19] It must be crowded in that unimaginable place, the centre of the earth – all the world's streams coming home. 'When the streames meet on all sides in the midst of the Earth they must needs be coarcted into a narrow roome & closely press together.'

Then again, perhaps an object's gravity was inherent, a quantity to be exactly measured, even if it varied from place to place: 'The gravity of a body in diverse places as at the top and bottom of a hill, in different latitudes &c. may be measured by an instrument' – he sketched a balance scale. He speculated about 'rays of gravity'. Then, *gravity* could also refer to a body's tendency to move, not downward, but in any direction; its tendency to remain in motion, once started. If such a tendency existed, no language yet had a word for it. Newton considered the problem of the cannonball, still rising, long after leaving the gun. 'Violent motion is made' – he struck the word *made* – 'continued either by the aire or by motion' – struck the word *motion* and replaced it with *force.*

Violent motion is ~~made~~ continued either by the aire or by ~~motion~~ force imprest or by the natural gravity in the body moved.

Yet how could the cannonball be helped along by the air? He noted that the air crowds more upon the front of a projectile than on the rear, '& must therefore rather hinder it'. So the continuing motion must come from some natural tendency in the object. But – *gravity?*

Some of his topics – for example, *Fluidity Stability Humidity Siccity*[20] – never progressed past a heading. No matter. He had set out his questions. *Of Heate & Cold. Atraction Magneticall. Colours. Sounds. Generation & Coruption. Memory.* They formed a programme, girded with measurements, clocks and scales, experiments both practical and imaginary. Its ambition encompassed the whole of nature.

One more mystery: *the Flux & Reflux of the Sea.* He considered a way to test whether the moon's 'pressing the atmosphere' causes the tides. Fill a tube with mercury or water; seal the top; 'the liquor will sink three or four inches below it leaving a vacuum (perhaps)'; then as the air is pressed by the moon, see if the water will rise or fall. He wondered whether the sea level rose by day and fell by night; whether it was higher in the morning or evening. Though fishermen and sailors around the globe had studied the tides for thousands of years, people had not amassed enough data to settle those questions.[21]

3

TO RESOLVE PROBLEMS
BY MOTION

Cambridge in 1664 had for the first time in its history a professor of mathematics, Isaac Barrow, another former Trinity College sizar, a decade older than Newton. Barrow had first studied Greek and theology; then left Cambridge, learned medicine, more theology, church history, and astronomy, and finally turned to geometry. Newton attended Barrow's first lectures. He was standing for examinations that year, on his way to being elected a scholar, and it was Barrow who examined him, mainly on the *Elements* of Euclid. He had not studied it before. At Stourbridge Fair he found a book of astrology and was brought up short by a diagram that required an understanding of trigonometry[1] – more than any Cambridge student was meant to know. He bought and borrowed more books. Before long, in a few texts, he had at hand a précis of the advanced mathematics available on the continent of Europe. He bought Franz van Schooten's *Miscellanies* and his Latin translation of Descartes's difficult masterpiece, *La Géométrie*; then William Oughtred's *Clavis Mathematicæ* and John Wallis's *Arithmetica Infinitorum*.[2] This reading remained far from comprehensive. He was inventing more than absorbing.

At the end of that year, just before the winter solstice, a comet appeared low in the sky, its mysterious tail blazing towards the west. Newton stayed outdoors night after night, noting a path against the background of the fixed stars, watching till it vanished in the light of each dawn, and only then returned to his room, sleepless and disordered. A comet was a frightening portent, a mutable and irregular traveller through the firmament. Nor was that all: rumours were reaching England of a new pestilence in Holland – perhaps from Italy or the Levant, perhaps from Crete or Cyprus.

Hard behind the rumours came the epidemic. Three men in London succumbed in a single house; by January the plague, this disease of population density, was spreading from parish to parish, hundreds dying each week, then thousands. Before the outbreak ran its course, in little more than a year, it killed one of every six Londoners.[3] Newton's mother wrote from Woolsthorpe:

> Isack
>
> received your leter and I perceive you
> letter from me with your cloth but
> none to you your sisters present thai
> love to you with my motherly lov
> you and prayers to god for you I
> your loving mother
>
> hanah
> wollstrup may the 6. 1665[4]

The colleges of Cambridge began shutting down. Fellows and students dispersed into the countryside.

Newton returned home. He built bookshelves and made a small study for himself. He opened the nearly blank thousand-page commonplace book he had inherited from his stepfather and named it his Waste Book.[5] He began filling it with reading notes. These mutated seamlessly into original research. He set himself problems; considered them obsessively; calculated answers, and asked new questions. He pushed past the frontier of knowledge (though he did not know this). The plague year was his transfiguration.[6] Solitary and almost incommunicado, he became the world's paramount mathematician.

Most of the numerical truths and methods that people had discovered, they had forgotten and rediscovered, again and again, in cultures far removed from one another. Mathematics was evergreen. One scion of *Homo sapiens* could still comprehend virtually all that the species knew collectively. Only recently had this form of knowledge begun to build upon itself.[7] Greek mathematics had almost vanished; for centuries, only Islamic mathematicians had kept it alive, meanwhile inventing abstract methods of problem solving called algebra. Now Europe became a special case: a region where people were using books and mail and a single language, Latin, to span tribal divisions across hundreds of miles; and where they were, self-consciously, receiving communications from a culture that had flourished and then disintegrated more than a thousand years before. The idea of knowledge as cumulative – a ladder, or a tower of stones, rising higher and higher – existed only as one possibility among many. For several hundred years, scholars of scholarship had considered that they might be like dwarfs seeing further by standing on the shoulders of giants, but they tended to believe more in rediscovery than in progress. Even now, when for the first time Western mathematics surpassed what had been known in Greece, many philosophers presumed they were merely uncovering ancient secrets, found in sunnier times and then lost or hidden.

With printed books had come a new metaphor for the world's organisation. The book was a container for information, designed in orderly patterns, encoding the real in symbols; so, perhaps, was nature itself. *The book of nature* became a favourite conceit of philosophers and poets: God had written; now we must read.[8] 'Philosophy is written in this grand book – I mean the universe – which stands continually open to our gaze,' said Galileo. 'But the book cannot be understood unless one first learns to comprehend the language and read the letters in which it is composed. It is written in the language of mathematics . . .'[9]

But by mathematics he did not mean numbers: 'Its characters are triangles, circles, and other geometrical figures, without which it is humanly impossible to understand a single word of it; without these, one is wandering about in a dark labyrinth.'

The study of different languages created an awareness of language: its arbitrariness, its changeability. As Newton learned Latin and Greek, he experimented with shorthand alphabets and phonetic writing, and when he entered Trinity College he wrote down a scheme for a 'universal' language, based on philosophical principles, to unite the nations of humanity. 'The Dialects of each Language being soe divers & arbitrary,' he declared, 'a generall Language cannot bee so fitly deduced from them as from the natures of things themselves.'[10] He understood language as a process, an act of transposition or translation – the conversion of reality into symbolic form. So was mathematics, symbolic translation at its purest.

For a lonely scholar seeking his own path through tangled thickets, mathematics had a particular virtue. When Newton got answers, he could usually judge whether they were right or wrong, no public disputation necessary. He read Euclid carefully now. The *Elements* – transmitted from ancient Alexandria via imperfect Greek copies, translated into medieval Arabic, and translated again into Latin – taught him the fundamental programme of deducing the properties of triangles, circles, lines and spheres from a few given axioms.[11] He absorbed Euclid's theorems for later use, but he was inspired by the leap of Descartes's *Géométrie*, a small and rambling text, the third and last appendix to his *Discours de la Méthode*.[12] This forever joined two great realms of thought, geometry and algebra. Algebra (a 'barbarous' art, Descartes said,[13] but it was his subject nonetheless) manipulated unknown quantities as if they were known, by assigning them symbols. Symbols recorded information, spared the memory, just as the printed book did.[14] Indeed, before texts could spread by printing, the development of symbolism had little point.

With symbols came equations: relations between quantities, and changeable relations at that. This was new territory, and Descartes exploited it. He treated one unknown as a spatial

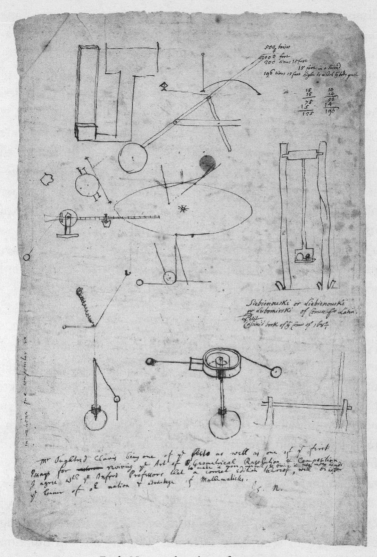

Early Newton drawings of apparatus.

dimension, a line; two unknowns thus define a plane. Line segments could now be added and even multiplied. Equations generated curves; curves embodied equations. Descartes opened the cage doors, freeing strange new bestiaries of curves, far more varied than the elegant conic sections studied by the Greeks. Newton immediately began expanding the possibilities, adding dimensions, generalising, mapping one plane to another with new coordinates. He taught himself to find real and complex roots of equations and to factor expressions of many terms – polynomials. When the infinite number of points in a curve correspond to the infinite solutions of its equation, then all the solutions can be seen at once, as a unity. Then equations have not just solutions but other properties: maxima and minima, tangents and areas. These were visualised, and they were named.

No one understands the mental faculty we call mathematical intuition; much less, genius. People's brains do not differ much, from one to the next, but numerical facility seems rarer, more special, than other talents. It has a threshold quality. In no other intellectual realm does the genius find so much common ground with the idiot savant. A mind turning inward from the world can see numbers as lustrous creatures; can find order in them, and magic; can know numbers as if personally. A mathematician, too, is a polyglot. A powerful source of creativity is a facility in translating, seeing how the same thing can be said in seemingly different ways. If one formulation doesn't work, try another.

Newton's patience was limitless. Truth, he said much later, was 'the offspring of silence and meditation'.[15]

And he said: 'I keep the subject constantly before me and wait 'till the first dawnings open slowly, by little and little, into a full and clear light.'[16]

Newton's Waste Book filled day by day with new research in this most abstract of realms. He computed obsessively. He worked out a way to transform equations from one set of axes to any alternative frame of reference. On one page he drew a hyperbola and set about calculating the area under it – 'squaring' it. He stepped past the algebra Descartes knew. He would not confine himself to expressions of a few (or many) terms; instead he constructed infinite series: expressions that continue forever.[17] An infinite series need not sum to infinity; rather, because the terms could grow smaller and smaller, they could close in on a goal or limit. He conceived such a series to square the hyperbola –

$$ax - \frac{x^2}{2} + \frac{x^3}{3a} - \frac{x^4}{4a^2} + \ldots$$

– and carried out the calculation to fifty-five decimal places: in all, more than two thousand tiny digits marching down a single page in orderly formation.[18] To conceive of infinite series and then learn to manipulate them was to transform the state of mathematics. Newton seemed now to possess a limitless ability to generalise, to move from one or a few particular known cases to the universe of all cases. Mathematicians had a glimmering notion of how to raise the sum of two quantities, $a + b$, to some power. Through infinite series,

Newton discovered in the winter of 1664 how to expand such sums to any power, integer or not: the general binomial expansion.

He relished the infinite, as Descartes had not. 'We should never enter into arguments about the infinite,' Descartes had written.

For since we are finite, it would be absurd for us to determine anything concerning the infinite; for this would be to attempt to limit it and grasp it. So we shall not bother to reply to those who ask if half an infinite line would itself be infinite, or whether an infinite number is odd or even, and so on. It seems that nobody has any business to think about such matters unless he regards his own mind as infinite.[19]

Yet it turns out that the human mind, though bounded in a nutshell, can discern the infinite and take its measure.

A special aspect of infinity troubled Newton; he returned to it again and again, turning it over, restating it with new definitions and symbols. It was the problem of the infinitesimal – the quantity, impossible and fantastic, smaller than any finite quantity, yet not as small as zero. The infinitesimal was anathema to Euclid and Aristotle. Nor was Newton altogether at ease with it.[20] First he thought in terms of 'indivisibles' – points which, when added to one another infinitely, could perhaps make up a finite length.[21] This caused paradoxes of dividing by zero:

Thus $\frac{2}{0}$ is double to $\frac{1}{0}$ & $\frac{0}{1}$ is double to $\frac{0}{2}$, for multiply the 2 first & divide the 2^{ds} by 0, & there results $\frac{2}{1}:\frac{1}{1}$ & $\frac{1}{1}:\frac{1}{2}\ldots$

– nonsensical results if o is truly zero, but necessary if o represents some indefinitely small, 'indivisible' quantity. Later he added an afterthought –

    (that is undetermined)
Tis indefinite $\wedge$ how greate a sphære may be made how greate a number may be reckoned, how far matter is divisible, how much time or extension wee can fansy but all the Extension that is, Eternity, $\frac{a}{0}$ are infinite.[22]

– blurring the words *indefinite* and *undetermined* by applying them alternately to mathematical quantities and degrees of knowledge. Descartes's reservations notwithstanding, the infinitude of the universe was in play – the boundlessness of

God's space and time. The infinitesimal – the almost nothing – was another matter. It might have been simply the inverse problem: the infinitely large and the infinitely small. A star of finite size, if it could be seen at an infinite distance, would appear infinitesimal. The terms in Newton's infinite series approached the infinitesimal. 'We are among infinities and indivisibles,' Galileo said, 'the former incomprehensible to our understanding by reason of their largeness, and the latter by their smallness.'[23]

Newton was seeking better methods – more general – for finding the slope of a curve at any particular point, as well as another quantity, related but once removed, the degree of curvature, rate of bending, 'the crookedness in lines'.[24] He applied himself to the tangent, the straight line that grazes the curve at any point, the straight line that the curve would *become* at that point, if it could be seen through an infinitely powerful microscope. He drew intricate constructions, more complex and more free than anything in Euclid or Descartes. Again and again he confronted the spectre of the infinitesimal: 'Then (if *hs* & *cd* have an infinitely little distance otherwise not) ...'; '...(which operacon cannot in this case bee understood to bee good unless infinite littleness may bee considered geometrically) ...'[25] He could not escape it, so he pressed it into service, employing a private symbol – a little o – for this quantity that was and was not zero. In some of his diagrams, two lengths differed 'but infinitely little', while two other lengths had 'no difference at all'. It was essential to preserve this uncanny distinction. It enabled him to find areas by infinitely partitioning curves and infinitely adding the partitions. He created 'a

Method whereby to square those crooked lines which may bee squared'[26] – to *integrate* (in the later language of the calculus).

As algebra melded with geometry, so did a physical counterpart, the problem of motion. Whatever else a curve was, it naturally represented the path of a moving point. The tangent represented the instantaneous direction of motion. An area could be generated by a line sweeping across the plane. To think that way was to think kinetically. It was here that the infinitesimal took hold. Motion was smooth, continuous, unbroken – how could it be otherwise? Matter might reduce to indivisible atoms, but to describe motion, mathematical points seemed more appropriate. A body on its way from a to b must surely pass through every point between. There must *be* points between, no matter how close a is to b; just as between any pair of numbers, more numbers must be found. But this continuum evoked another form of paradox, as Greek philosophers had seen two thousand years before: the paradox of Achilles and the tortoise. The tortoise has a head start. Achilles can run faster but can never catch up, because each time he reaches the tortoise's last position, the tortoise has managed to crawl a bit further ahead. By this logic Zeno proved that no moving body could ever reach any given place – that motion itself did not exist. Only by embracing the infinite and the infinitesimal, together, could these paradoxes be banished. A philosopher had to find the sum of infinitely many, increasingly small intervals. Newton wrestled with this as a problem of words: swifter, slower; least distance, least progression; instant, interval.

That it may be knowne how motion is swifter or slower consider: that there is a least distance, a least progression in motion & a least degree of time . . . In each degree of time wherein a thing moves there will be motion or else in all those degrees put together there will be none: . . . no motion is done in an instant or intervall of time.[27]

A culture lacking technologies of time and speed also lacked basic concepts that a mathematician needed to quantify motion. The English language was just beginning to adopt its first unit of velocity: the term *knot*, based on the sailor's only speed-measuring device, the log line heaved into the sea. The science most eager to understand the motion of earthly objects, ballistics, measured the angles of gun barrels and the distances their balls travelled, but scarcely conceived of velocity; even when they could define this quantity, as a ratio of distance and time, they could not measure it. Galileo, when he dropped weights from towers, could make only the crudest estimates of their velocity, though he used an esoteric unit of time: *seconds* of an hour. Newton was struck by the ambition in his exactitude: 'According to Galileo an iron ball of 100 lb. Florentine (that is 78 lb. at London avoirdupois weight) descends 100 Florentine braces or cubits (or 49.01 Ells, perhaps 66 yds.) in 5 seconds of an hour.'[28]

In the autumn of 1665 he made notes on 'mechanical' lines, as distinguished from the merely geometric. Mechanical curves were those generated by the motion of a point, or by two such motions compounded: spirals, ellipses and cycloids. Descartes had considered the cycloid, the curve generated by a point on a circle as the circle rolls along a line. He regarded this oddity as suspect and unmathematical, because it could not (before the calculus) be described analytically. But such artefacts from the new realm of mechanics kept intruding on mathematics. Hanging cables or sails in the wind traced mechanical curves.[29] If a cycloid was mechanical, it was nevertheless an abstraction: a creature of several motions, or rates, summed in a certain way. Indeed, Newton now saw ellipses in different lights – geometrical and analytical. The ellipse was the effect of a quadratic formula. Or it was the closed line drawn in the dirt by the 'gardener's' construction, in which a loose cord is tied to two pegs in the ground: 'keeping it so stretched out draw the point b about & it shall describe the Ellipsis'.[30] Or it was a circle with extra freedom; a circle with one constraint removed; a squashed circle, its centre bifurcating into a pair of foci. He devised procedures for drawing tangents to mechanical curves, thus measuring their slopes; and, in November, proposed a method for deducing, from two or more such lines, the corresponding relation between the velocities of two or more moving bodies.[31]

He found tangents by computing the relationship between points on a curve separated by an infinitesimal distance. In the computation, the points almost merge into one, 'conjoyne, which will happen when $bc = 0$, vanisheth into nothing'.[32] That 0 was an artifice, a gadget for the infinitesimal, as an arbitrarily small increment or a moment of time. He showed how the terms with 0 'may be ever blotted out'.[33] Extending his methods, he also quantified rates of bending, by finding centres of curvature and radii of curvature.

A geometrical task matched a kinetic task: to measure curvature was to find a rate of change. *Rate of change* was itself an abstraction of an abstraction; what velocity was to position, acceleration was to velocity. It was differentiation (in the later language of the calculus). Newton saw this system whole: that problems of tangents were the inverse of problems of quadrature; that differentiation and integration are the same act, inverted. The procedures seem alien, one from the other, but what one does, the other undoes. That is the fundamental theorem of the calculus, the piece of mathematics that became essential knowledge for building engines and measuring dynamics. Time and space – joined. *Speed* and *area* – two abstractions, seemingly disjoint, revealed as cognate.

Repeatedly he started a new page – in November 1665, in May 1666 and in October 1666 – in order to essay a system of propositions needed '*to resolve Problems by motion*'.[34] On his last attempt he produced a tract of twenty-four pages, on eight sheets of paper folded and stitched together. He considered points moving towards the centres of circles; points moving parallel to one another; points moving 'angularly' or 'circularly' – this language was unsettled – and points moving along lines that intersected planes. A variable representing time underlay his equations – time as an absolute background for motion. When velocity changed, he imagined it changing smoothly and continuously – across infinitesimal moments, represented by that o. He issued himself instructions:

Set all the termes on one side of the Equation that they become equall to nothing. And first multiply each terme by so many times

$\frac{p}{x}$ as x hath dimensions in the terme. Secondly multiply each term by so many times $\frac{q}{y}$... & if there bee still more unknowne quantitys doe like to every unknowne quantity.[35]

Time was a flowing thing. In terms of velocity, position was a function of time. But in terms of acceleration, velocity was itself a function of time. Newton made up his own notation, with combinations of superscript dots, and vocabulary, calling these functions 'fluents' and 'fluxions': flowing quantities and rates of change. He wrote it all several times but never quite finished.

In creating this mathematics Newton embraced a paradox. He believed in a discrete universe. He believed in atoms, small but ultimately indivisible – *not* infinitesimal. Yet he built a mathematical framework that was not discrete but continuous, based on a geometry of lines and smoothly changing curves. 'All is flux, nothing stays still,' Heraclitus had said two millennia before. 'Nothing endures but change.' But this state of being – in *flow*, in *change* – defied mathematics then and afterwards. Philosophers could barely observe continuous change, much less classify it and gauge it, until now. It was nature's destiny now to be mathematised. Henceforth space would have dimension and measure; motion would be subject to geometry.[36]

Far away across the country multitudes were dying in fire and plague. Numerologists had warned that 1666 would be the Year of the Beast. Most of London lay in black ruins: fire had begun in a bakery, spread in the dry wind across thatch-roofed houses, and blazed out of control for four days and four nights.

The new king, Charles II – having survived his father's beheading and his own fugitive years, and having outlasted the Lord Protector, Cromwell – fled London with his court. Here at Woolsthorpe the night was strewn with stars, the moon cast its light through the apple trees, and the day's sun and shadows carved their familiar pathways across the wall. Newton understood now: the projection of curves onto flat planes; the angles in three dimensions, changing slightly each day. He saw an orderly landscape. Its inhabitants were not static objects; they were patterns, process and change.

What he wrote, he wrote for himself alone. He had no reason to tell anyone. He was twenty-four and he had made tools.

4

TWO GREAT ORBS

Historians came to see Newton as an end-point: the 'culmination' and 'climax' of an episode in human affairs conventionally called the Scientific Revolution. Then that term began to require apologies or ironic quotation marks.[1] Ambivalence is appropriate, when one speaks of the turning point in the development of human culture, the time when reason triumphed over unreason. The Scientific Revolution is a story, a narrative frame laid down with hindsight. Yet it exists and existed, not just in the backward vision of historians but in a self-consciousness among a small number of people in England and Europe in the seventeenth century. They were, as they thought, virtuosi. They saw something new in the domain of knowledge; they tried to express the newness; they invented academies and societies and opened channels of communication to promote their break with the past, their *new* science.

We call the Scientific Revolution an epidemic, spreading across the continent of Europe during two centuries: 'It would come to rest in England, in the person of Isaac Newton,' said the physicist David Goodstein. 'On the way north, however, it stopped briefly in France . . .'[2] Or a relay race, run by a team of heroes who passed the baton from one to the next: COPERNI-CUS to KEPLER to GALILEO to NEWTON. Or the overthrow and destruction of the Aristotelian cosmology: a world-view that staggered under the assaults of Galileo and Descartes and finally expired in 1687, when Newton published a book.[3]

For so long the earth had seemed the centre of all things. The constellations turned round in their regular procession. Just a few bright objects caused a puzzle – the planets, wanderers, like gods or messengers, moving irregularly against the fixed backdrop of stars. In 1543, just before his death, Nicolaus Copernicus, Polish astronomer, astrologer and mathematician, published the great book *De Revolutionibus Orbium Coelestium* ('On the Revolutions of the Heavenly Spheres'). In it he gave order to the planets' paths, resolving them into perfect circles; he set the earth in motion and placed an immobile sun at the centre of the universe.[4]

Johannes Kepler, looking for more order in a growing thicket of data, thousands of painstakingly recorded observations, declared that the planets could not be moving in circles. He suspected the special curves known to the ancients as ellipses. Having thus overthrown one kind of celestial perfection, he sought new kinds, believing fervently in a universe built on geometrical harmony. He found an elegant link between geometry and motion by asserting that an imaginary line from a planet to the sun sweeps across equal areas in equal times.[5]

Galileo Galilei took spyglasses – made by inserting spectacle makers' lenses into a hollow tube – and pointed them upwards towards the night sky. What he saw both inspired and disturbed him: moons orbiting Jupiter; spots marring the sun's flawless face; stars that had never been seen – 'in numbers ten times exceeding the old and familiar stars'.[6] He learned, 'with all the certainty of sense evidence', that the moon 'is not robed in a smooth polished surface but is in fact rough and uneven'. It has mountains, valleys and chasms. (He also thought he had detected an atmosphere of dense and luminous vapours.)

He took pains to detail an unfamiliar fact of arithmetic: that, because in his spyglass the moon's diameter appeared thirty times larger, its apparent area was magnified by 900 and its apparent volume by 27,000 – a square law and a cube law. This was essentially the only mathematics in his report, *The Starry Messenger.*[7]

It was strange to think of these dots of light as worlds, and more strange to think of a world – the *whole* world – as a body in motion, comparable to a mere stone. Yet without understanding motion, no one could place the heavenly bodies. There could be no cosmology without dynamics. Galileo felt this. What he saw in the skies of Florence in 1610, English pamphleteers tried to convey a generation later. In London a young chaplain, John Wilkins, began writing anonymous screeds. First, in 1638, *The Discovery of a New World; or, a Discourse tending to prove, that it is probable there may be another habitable World in the Moon.*[8]

Among all the celestial mysteries, the moon was special – so near, so changeling, so portentous. It stirred madness in weak minds; people were known to grow lunatic on a monthly cycle. Empedocles saw the moon as 'a globe of pure congealed air, like hail inclosed in a sphere of fire'. Aristotle held it to be solid and opaque, whereas Julius Caesar said it must be transparent and pure, of the same essence as the heavens. Plain observation, night after night, failed to settle such matters. 'You may as soon persuade some country peasants that the moon is made of green cheese (as we say) as that it is bigger than his cart-wheel,' wrote Wilkins, 'since both seem equally to contradict his sight, and he has not reason enough to lead him farther than his senses.'[9]

How far could reason lead, without help? Francis Bacon, who had practised logic and disputation as the king's Learned Counsel and Attorney-General, lamented a natural philosophy built solely on words, ostentation, the elaborate knitting together of established ideas.

All the philosophy of nature which is now received, is either the philosophy of the Grecians, or that other of the alchemists . . . The one is gathered out of a few vulgar observations, and the other out of a few experiments of a furnace. The one never faileth to multiply words, and the other ever faileth to multiply gold.[10]

He argued for experiment – the devising of 'Crucial Instances' to divide the true from the false. Was the moon flame-like and airy or solid and dense? Since the moon reflects the sun's light, Bacon proposed, a crucial instance would be a demonstration that a flame or other rare body does or does not reflect light. Perhaps the moon also 'raises the waters', Bacon suggested, and 'makes moist things swell'. He proposed to call this effect Magnetic Motion.[11]

Wilkins cited the lunar observations of many authorities:

Herodotus, the Venerable Bede, the Romish divines, the Stoics, Moses and Thomas Aquinas. But at last he chose a new witness.

I shall most insist on the observation of Galilæus, the inventor of that famous perspective, whereby we may discern the heavens hard by us; whereby those things which others have formerly guessed at, are manifested to the eye, and plainly discovered beyond exception or doubt.[12]

With his glass, Galileo could see plainly at a distance of sixteen miles what the naked eye could scarcely see at a mile and a half. He saw mountains and valleys; he saw a sphere of thick vaporous air; from these it was but a short step to infer wind and rain, seasons and weather, and so, Wilkins concluded, inhabitants. 'Of what kind they are, is uncertain,' he conceded. 'But I think that future ages will discover more; and our posterity, perhaps, may invent some means for our better acquaintance with these inhabitants.' As soon as the art of flying is discovered, he said, we should manage to transplant colonies to that other world. After all, time is the father of truth; ages passed before men crossed the seas and found other men at the far side of the world; surely other excellent mysteries remain to be discovered.

Wilkins urged that the strangeness of his opinions should be no reason to reject them. The surprising discovery of another New World weighed heavily: 'How did the incredulous gaze at Columbus, when he promised to discover another part of the earth?'

Still, he agreed that the idea of multiple worlds brought paradoxical difficulties. The most troublesome was the tendency of heavy bodies to fall down: their *gravity*. 'What a huddling and confusion must there be, if there were two places for gravity and two places for lightness?'[13] Which way should bodies of that other world fall? To where should its air and fire ascend? Can we expect pieces of the moon to fall to earth?

He answered these questions in the terms of Copernicus and Kepler: by proposing that two worlds must have two centres of gravity. 'There is no more danger of their falling into our world, than there is fear of our falling into the moon.' He reminded his readers of the simple nature of gravity: 'nothing else, but such a quality as causes a propension in its subject to tend downwards towards its own centre'.[14]

The discovery of new worlds had lit a fuse leading to the destruction of the Aristotelian conception of gravity. It was inevitable. A multitude of worlds implied a multitude of reference frames. *Up* and *down* became relative terms, in the imaginations of philosophers, contrary to common experience. Wilkins did not shrink from considering the problem of what would happen to an object – a bullet, perhaps – sent to such a great height that it might depart 'that magnetical globe to which it did belong'. It might just come to rest, he decided. Outside the earth's sphere of influence, pieces of earth should lose their gravity, or their susceptibility to gravity. He offered a 'similitude':

As any light body (suppose the sun) does send forth its beams in an orbicular form; so likewise any magnetical body, for instance a

round loadstone, does cast abroad his magnetical vigour in a sphere
. . . Any other body that is like affected coming within this sphere
will presently descend towards the centre of it, and in that respect
may be styled heavy. But place it without this sphere, and then the
desire of union ceaseth, and so consequently the motion also.[15]

Newton read Wilkins as a boy in Grantham, at the apothecary Clarke's.[16] Whatever else he thought about the moon,
he knew it was a great planetary object travelling through
space at high speed. The mystery was why. Carried along,
as Descartes said, in a vortex? Newton knew how big the
moon was and how far away. By virtue of a coincidence, the
moon's apparent size was almost exactly the same as the sun's,
about one-half degree of arc, the coincidence that makes a
solar eclipse such a perfect spectacle. It was necessary now to
forge mental links across many orders of magnitude in scale:
between the everyday and the unimaginably vast. Sitting
in the orchard behind his farmhouse, musing continually on
geometry, Newton could see other globes, dangling from their
stems. A two-inch apple at a distance of twenty feet subtended
the same half-degree in the sky. These ratios were second
nature now, the congruent Euclidean triangles inscribed in his
mind's eye. When he thought about the magnitude of these
bodies, another automatic part of the picture was an inverse
square law: something varies as $1/x^2$. A disc twice as far away
would seem not one-half as bright but one-fourth.

Newton was eager, as the Greeks had not been, to extend
the harmony and abstraction of mathematics to the crude sublunary world in which he lived. An apple was no sphere, but

he understood it to be flying through space along with the rest of the earth's contents, spinning across 25,000 miles each day. Why, then, did it hang gently downward, instead of being flung outward like a stone whirled around on a string? The same question applied to the moon: what pushed it or pulled it away from a straight path?

Many years later Newton told at least four people that he had been inspired by an apple in his Woolsthorpe garden – perhaps an apple actually falling from a tree, perhaps not. He never wrote of an apple. He recalled only:

I began to think of gravity extending to the orb of the Moon . . .

– gravity as a force, then, with an extended field of influence; no cut-off or boundary –

& computed the force requisite to keep the Moon in her Orb with the force of gravity at the surface of the earth . . . & found them answer pretty nearly. All this was in the two plague years of 1665–1666. For in those days I was in the prime of my age for invention & minded Mathematicks and Philosophy more than at any time since.[17]

Voltaire did mention the apple, as did other memoirists, and their second- and third-hand accounts gradually formed the single most enduring legend in the annals of scientific discovery.[18] And the most misunderstood: Newton did not need an apple to remind him that objects fell to earth. Galileo had not only seen objects fall but had dropped them from towers

and rolled them down ramps. He had grasped their acceleration and struggled to measure it. But most emphatically he
declined to explain it. 'The present does not seem to be
the proper time to investigate the cause of the acceleration,'
Galileo wrote, '. . . [but] merely to investigate and to demonstrate some of the properties of accelerated motion (whatever
the cause of this acceleration may be).'[19]

Nor did Newton comprehend universal gravitation in a
flash of insight. In 1666 he was barely beginning to understand.
What he suspected about gravity he kept private for decades
to come.

The apple was nothing in itself. It was half of a couple –
the moon's impish twin. As an apple falls towards the earth, so
does the moon: falling away from a straight line, falling around
the earth. Apple and moon were a coincidence, a generalisation, a leap across scales, from close to far and from ordinary
to immense. In his study and in his garden, in his state of
incessant lonely contemplation, his mind alive with new
modes of geometry and analysis, Newton made connections
between distant realms of thought. Still, he was unsure. His
computations were ambiguous; he only *found them answer pretty
nearly*. He was attempting rare exactitude, more than any
available raw data could support. Even the units of measure
were too crude and variable. He took the mile to be 5,000
feet.[20] He set one degree of the earth's latitude at the equator
equal to sixty miles, an error of about 15 per cent. Some units
were English, some antique Latin, others Italian: mile, passus,
brace, pedes. He came up with a datum for the speed of the
revolving earth: 16,500,000 cubits in six hours.[21] He struggled

to arrive at a datum for the rate of fall due to gravity. He had Galileo's calculations, in a new translation: one hundred cubits in five seconds.[22] He tried to derive his own measurements using a weight hanging on a cord and swinging in circles – a conical pendulum. This needed patience. He noted the pendulum making 1,512 'ticks' in an hour.[23] He arrived at a constant for gravity more than double Galileo's. He concluded that a body on the earth's surface is drawn downwards by gravity 350 times stronger than the tendency of the earth's rotation to fling it outwards.

To make the arithmetic work at all, he had to suppose that the power of attraction diminished rapidly according to distance from the centre of the earth. Galileo had said that bodies fall with constant acceleration, no matter how far they are from the earth; Newton sensed that this must be wrong. And it would not be enough for gravity to fade in proportion to distance. He estimated that the earth attracted an apple 4,000 times as powerfully as it attracted the distant moon. If the ratio – like brightness, and like apparent area – depended on the *square* of distance, that might answer pretty nearly.[24]

He reckoned the distance of the moon at sixty times the earth's radius; if the moon were sixty times further than the surface of the earth from the centre of the earth, then the earth's gravity might be 3,600 times weaker there. He also derived this inverse-square law by an inspired argument from an observation of Kepler's: the time a planet takes to make one orbit grows as the 3/2 power of its distance from the sun.[25] Yet, with the data he had, he could not quite make the numbers work. He still found it necessary to attribute some of the moon's motion to the vortices of Descartes.

He needed new principles of motion and force. He had tried some out in the *Questiones*, and now, in the plague year, he tried again. He wrote 'axioms' in the Waste Book:

1. If a quantity once move it will never rest unlesse hindered by some externall caus.

2. A quantity will always move on in the same streight line (not changing the determination nor celerity of its motion) unless some externall cause divert it.[26]

Thus circular motion – orbital motion – demanded explanation. So far, the *external cause* was missing from the picture. And Newton posed himself a challenge: it ought to be possible to quantify this cause.

3. There is exactly so much required so much and noe more force to reduce a body to rest as there was to put it upon motion.

He continued through dozens more axioms, comprising a logical whole, but a tangled one. He was hampered by the chaos of language – words still vaguely defined and words not quite existing. He conceived of *force* as a thing to be measured – but in what units? Was force inherent in bodies, as Descartes thought? Or was force an external agent, impinging on bodies and changing a differently named quantity: *quantity of motion*, or *quantity of mutation in its state*[27]; or *whole motion*, or *force of motion*? Whatever this missing concept was, it differed from velocity and direction. Axiom 100:

A body once moved will always keepe the same celerity, quantity and determination of its motion.[28]

At twenty-four, Newton believed he could marshal a complete science of motion, if only he could find the appropriate lexicon, if only he could set words in the correct order. Writing mathematics, he could invent his own symbols and form them into a mosaic. Writing in English, he was constrained by the language at hand.[29] At times his frustration was palpable in the stream of words. Axiom 103:

. . . as the body (*a*) is to the body (*b*) so must the power or efficacy vigor strength or virtue of the cause which begets the same quantity of velocity . . .[30]

Power efficacy vigor strength virtue – something was missing. But these were the laws of motion, in utero.

5

BODYS & SENSES

He was looking inward as well as outward. Introspection told him that his imagination could see things as they really were. 'Phantasie is helped,' he noted, 'by good aire fasting moderate wine.' But it is also 'spoiled by drunkenesse, Gluttony, too much study'. He added: from too much study, and from extreme passion, 'cometh madnesse'.[1]

He wished to understand light itself – but did light's essence lie outside or within the soul of the observer? In all the blooming perplexity of new philosophy, little was as muddled as the boundary between the perceived and the perceiver. Surely the mind, composed of pure thought, must have a point of contact with the body – at the pineal gland, Descartes proposed. The poet Andrew Marvell, graduate of Trinity College and now Member of Parliament for Hull, imagined the body and soul as enslaved, each by the other: 'A soul hung up, as 'twere, in chains of nerves and arteries and veins.'[2] For Aristotle optics had been first a science not of light but of sight.

Newton, in his *Questiones*, had pondered the difficulty of understanding the senses, when those very senses were employed as the agents of understanding.

The nature of things is more securely & naturally deduced from their operations out upon another than upon our senses. And when by the former experiments we have found the nature of bodys, by the latter we may more clearly find the nature of our senses. But so long as we are ignorant of the nature of both soul and body we cannot clearly distinguish how far an act of sensation proceeds from the soul and how far from the body.[3]

With this paradox in mind, Newton, experimental philosopher, slid a bodkin into his eye socket between eyeball and bone. He pressed with the tip until he saw 'severall white darke & coloured circles . . . Which circles were plainest when I continued to rub my eye with the point of the bodkin.' Yet when he held both eye and bodkin still, the circles would begin to fade.[4] Was light a manifestation of pressure, then?

Almost as recklessly, he stared with one eye at the sun, reflected in a looking glass, for as long as he could bear. He sensed that colour – perhaps more than any of the other *qualities* of things – depends on 'imagination and fantasy and invention'.[5] He looked away at a dark wall and saw circles of colour. There was a 'motion of spirits' in his eye. These slowly decayed and finally vanished. Were they real or phantasm? Could such colours ever be *real*, like the colours he had learned to make from crushed berries or sheep's blood? After looking at the sun, he seemed to perceive light objects as red and dark objects as blue. Strangely, he found that he could reproduce these effects, with practice, by pure, wilful thought. 'As often as I went into the dark & intended my mind upon them as when a man looks earnestly to see any thing which is difficult to be seen, I could make the phantasm return without looking any more upon the sun.'[6] He repeated the experiment until he began to fear permanent damage and shut himself up in a dark room. He remained there for three days; only then did his sight begin to clear.

Experiment – observation – *science*: these modern words were impressing themselves upon him. He read them in a new book from London, titled *Micrographia*: 'The Science of Nature has been already too long made only a work of the Brain and the Fancy. It is now high time that it should return to the plainness and soundness of Observations on material and obvious things.'[7] The author was Robert Hooke, a brilliant and ambitious man seven years Newton's senior, who wielded the microscope just as Galileo had the telescope. These were the instruments that penetrated the barrier of scale and opened a view into the countries of the very large and the very small. Wonders were revealed there. The old world – the world of ordinary scales – shrank into its place in a continuum, one order among many. Like Galileo, Hooke made meticulous drawings of strange new sights and popularised his instrument as a curiosity for wealthy aristocrats – though, after they bought the device from the lens shop in London where he sometimes worked, they rarely succeeded in seeing anything but vague shadows. Hooke was Newton's inspiration now (though Newton never acknowledged that). He became Newton's goad, nemesis, tormentor, and victim.

Hooke had a unique post. He was employed, if seldom actually paid, as Curator of Experiments to a small group of men who formed, in 1662, what they called the Royal Society of London. They meant to be a new sort of institution: a national society dedicated to promoting – and especially 'communicating' – what they called 'the New Philosophy' or 'Experimental Philosophy'.[8] Amazing discoveries warranted this banner: comets and new stars; the circulation of the blood; the grinding of glasses for telescopes; the possibility of vacuities (and nature's abhorrence thereof); the descent of heavy bodies; and diverse other things.[9]

Nullius in verba was the Royal Society's motto. Don't take anyone's word for it.[10] These gentlemen had begged for and received the king's patronage, but patronage meant good will only; the society collected from its members a shilling at a time and strained to find meeting places. Among the founders was John Wilkins, author of *The Discovery of a New World* a generation before. If one man was their muse, he was the late Francis Bacon, who had written:

We must . . . completely resolve and separate Nature, not by fire, certainly, but by the mind, which is a kind of divine fire . . . There will remain, all volatile opinions vanishing into smoke, the affirmative form, solid, true and well-defined. Now this is quickly said, but it is only reached after many twists and turns.[11]

The twists and turns became the responsibility of the Curator of Experiments, Hooke, technician and impresario. He demonstrated experiments with air pumps. At one meeting he cut open the thorax and belly of a living dog, observed its beating heart, and used a bellows to inflate its lungs in an experiment on respiration, which he later felt reluctant to repeat 'because of the torture of the creature'.[12] Another meeting dazzled and confused the Duchess of Newcastle with colours, magnets, microscopes, roasted mutton and blood.[13] This was all science, a new spirit and almost a method: persuasion from practical experience, and formalised recording of data. Hooke lacked mathematics but not ingenuity. He invented or improved barometers, thermometers and wind gauges, and he tracked London weather obsessively.[14]

In *Micrographia* he displayed the 'new visible world' to be seen though the instrument he described as an artificial organ. 'By the help of Microscopes, there is nothing so small, as to escape our inquiry,' he declared.[15] As a geometer begins with a mathematical point, he examined the point of a needle – perfectly sharp, yet under the microscope, blunt and irregular. By analogy he suggested that the earth itself, seen from a great enough distance, would shrink to a scarcely visible speck. More specks were to be found in printed books: he proceeded to study and draw the mark of a full stop, the punctuation mark – again surprisingly rough and irregular, 'like a great splatch of London dirt'.[16] He found wonderment in the edge of a razor and the weft of fine linen. He discovered shifting, iridescent colours in thin flakes of glass. He knew that Descartes had seen a rainbow of colours in light passed through a prism or a water drop, and he compared microscopic rainbows.

And here he made his book something more than a registry and gazetteer for his new world. He notified readers that he offered a *theory* – a complete and methodical explanation of light and colour. Aristotle had thought of colour as a commingling of black and white. His followers considered colours fundamental qualities of matter, carried by light into the eye. Descartes had speculated that colour came from globules of light changing speed when refracted by glass or water. Hooke disputed this and, grandly invoking the shade of Bacon, turned to experiment: an '*Experimentum Crucis*, serving as Guide or Land-mark'.[17] True, Hooke observed, a prism produces colours when refracting light. But he asserted that refraction was not necessary. His landmark was the production of colour in transparent substances:[18] 'for we find, that the Light in the open Air, either in or out of the Sun-beams, and within a Room, either from one or many Windows, produces much the same effect.'

Light is born of motion, he argued. 'That all kind of *fiery burning Bodies* have their parts in motion, I think will be very easily granted me.' Sensing more than he could truly see, he asserted that all luminous bodies are in motion, perhaps vibrating: sparks, rotting wood, and fish. Further, he observed, or thought he observed, that two colours were fundamental: blue and red. They were caused by 'an impression on the retina of an oblique and confus'd pulse of light'.[19] Where red and blue 'meet and cross each other', the imperfection generated 'all kinds of greens'. And here his theory ended. 'It would be somewhat too long a work for this place zetetically to examine, and positively to prove, what particular kind of motion it is . . . It would be too long, I say, here to insert the discursive progress by which I inquir'd after the properties of the motion of Light . . .'[20] Yet all in all he claimed to have explained everything; to have given – 'newly' given – the causes

capable of explicating all the Phenomena of colours, not onely of those appearing in the Prisme, Water-drop, or Rainbow . . . but of all that are in the world, whether they be fluid or solid bodies, whether in thick or thin, whether transparent, or seemingly opacous.[21]

Newton absorbed this bold claim.[22] He had no microscope and no chance of obtaining one. For that matter, he had no room with more than one window. He did have a prism. He darkened his study and made a hole in the window shutter to let in a sunbeam, white light, the purest light, light with no intrinsic colour, philosophers still thought. He performed his own experiments – even, he felt, an *experimentum crucis*. He noted the results and told no one.

Bacon had also warned: 'God forbid that we should give out a dream of our own imagination for a pattern of the world.'[23]

The plague abating, Newton returned to Cambridge, where among those he did not tell of his experiments was the pro fessor of mathematics, Isaac Barrow.

6

THE ODDEST IF NOT THE MOST
CONSIDERABLE DETECTION

Newton's status at Trinity improved. In October 1667 the college elected fellows for the first time in three years: men entitled to wages (two pounds a year), a room, continuing membership in the academic community, and the use of the library. Each new fellow swore: 'I will embrace the true religion of Christ with all my soul . . . I will either set Theology as the object of my studies and will take holy orders when the time prescribed by these statutes arrives, or I will resign from the college.' Chastity was expected and marriage forbidden. Newton bought shoes and cloth for the gown of a bachelor of arts. Besides his stipend he received small sums from his mother and (very rarely) from pupils he tutored. He bought a set of old books on alchemy, along with glasses, a tin furnace, and chemicals: aqua fortis, sublimate, vinegar, white lead, salt of tartar.[2] With these he embarked on a programme of research more secret than ever.

But he also continued his mathematical investigations, and he shared some of these with Barrow. He began to list cubic equations: curves in three dimensions, more various and complex than the ellipses and hyperbolas of two-dimensional mathematics. He attacked this subject as a classifier, trying to sort all such curves into species and subspecies.[3] As he had done with the calculus, he approached this analytic geometry from two directions at once: from the perspective of algebra, where cubic equations begin with the form $x^3 + ax^2 + bx + c = 0$; and from a kinematic perspective, describing these creatures in terms of their construction, as the results of points and curves moving through space. He plotted in his notebooks fifty-eight distinct species of cubics. He sought ever greater generality.

Barrow showed him a new book from London, *Logarithmotechnia*, by Nicholas Mercator, a mathematics tutor and member of the Royal Society. It presented a method of calculating logarithms from infinite series and thus gave Newton a shock: his own discoveries, rediscovered. Mercator had constructed an entire book – a useful book, at that – from a few infinite series. For Newton these were merely special cases of the powerful approach to infinite series he had worked out at Woolsthorpe. Provoked, he revealed to Barrow a bit more of what he knew. He drafted a paper in Latin, 'On Analysis by Infinite Series'. He also let Barrow post this to another Royal Society colleague, a mathematician, John Collins,[4] but he insisted on anonymity. Only after Collins responded enthusiastically did he let Barrow identify him: 'I am glad my friends paper giveth you so much satisfaction. his name is Mr Newton; a fellow of our College, & very young . . . but of an extraordinary genius and proficiency in these things.'[5] It was the first transmission of Newton's name south of Cambridge.

At long distance, in messages separated by days or by months, Newton and Collins now engaged in a dance. Newton teased Collins with tantalising fragments of mathematical insight. Collins begged for more. Newton delayed and withdrew. A table resolving equations of three dimensions was 'pretty easy and obvious enough', he declared. 'But I cannot perswade my selfe to undertake the drudgery of making it.'[6] Collins bruited some of Newton's handiwork to several other mathematicians, in Scotland, France and Italy. He sent books to Newton and posed questions: for example, how to calculate the rate of interest on an annuity. Newton sent a formula for that but insisted that his name be withheld if Collins published it: 'For I see not what there is desirable in publick esteeme, were I able to acquire & maintaine it. It would perhaps increase my acquaintance, the thing which I cheifly study to decline.'[7] Nonetheless his name was being whispered. James Gregory, the Scots mathematician, heard it. He was struggling with an unsolved problem of analytic geometry that he read in new lectures by Barrow. 'I despaire of it my self, and therfor I doe humblie desire it of any els who can resolve it,' he wrote Collins. 'I long to see that peece of Mr Newton which is generallie applied to al curvs.'[8]

When Barrow prepared his lectures for publication, he asked Newton to help him edit the manuscripts, particularly his *Optical Lectures*.[9] These appeared in 1669, with Barrow's effusive acknowledgement of 'a Man of great Learning and Sagacity, who revised my Copy and noted such things as wanted correction'. Yet Newton knew what Barrow did not: that the whole project wanted correction. Barrow imagined that colour had something to do with compression and rarification and excitation of light; that red might be 'broken and interrupted by shadowy interstices' while blue involved 'white and black particles arranged alternately'.[10] Barrow's protégé had already done private research that rendered these optics obsolete. Anyway, Barrow had ambitions elsewhere. He was a favourite of the king, hoped for advancement, and thought of himself more as a theologian than a mathematician. Before the end of the year, he resigned his post as Lucasian professor, yielding it to Newton, twenty-seven years old.[11]

The young professor gained relative security. He could be removed only for serious crime; the statutes specified fornication, heresy and voluntary manslaughter.[12] He was expected to read a lecture on mathematics (broadly construed) each week during the academic term and deposit a copy in the university library. But he disregarded this obligation far more than he fulfilled it. When he did lecture, students were scarce. Sometimes he read to a bare room or gave up and walked back to his chambers.[13] The existence of this new professorship reflected a sense that mathematics was an art useful to the growing nation – its architects, tradesmen and sailors – but cubic curves and infinite series had no use in a trade or on a ship. Such mysteries were as recondite as the researches Newton was beginning to undertake alone in his chambers with his tin crucible.

Instead of mathematics he chose to lecture on light and colour. The invention of telescopes had spurred intense interest in the properties of light, he noted, yet the geometers had 'hitherto erred'. So he proposed to add his own discoveries 'to what my reverend predecessor last delivered from this Place'.[14] He considered the phenomenon of refraction, the bending of light when it passes from one medium to another, as from air to glass (lenses being the offspring of refraction and geometry). Wearing a professor's gown of scarlet, he stood before the few students who attended and delivered news: rays of coloured light differ from one another in how sharply they are refracted. Each colour has its own degree of refraction. This was a bare, mathematical claim, with none of the romance or metaphor that usually ornamented the philosophy of light.

Newton was not just drawing and calculating; he was also grinding glass and polishing lenses in difficult, nonspherical curves. Telescope makers had learned to their sorrow that spherical lenses blurred their images, inevitably, because rays of light failed to meet at a single point. Also, the larger they made the lenses, the more they saw rings of unwanted colour – and Newton understood these now. The problem lay not in imperfect craft but in the very nature of white light: not simple but complex; not pure but mixed; *a heterogeneous mixture of differently refrangible rays.*[15] Lenses were after all prisms at their edges. He tried a new kind of telescope, based on a reflecting mirror instead of a refracting lens.[16] A big mirror would gather more light than a small lens – in proportion to its area, or to the square of its diameter. The difficulty was a matter of craft: how to polish metal to the smoothness of glass. With his furnace and putty and pitch he cast a tin and copper alloy and refined its surface, grinding with all his strength. In 1669 he had a stubby little tube six inches long and magnifying forty times – as much as the best telescopes in London and Italy, and as much as a refracting telescope ten times longer.[17] He kept it for two years. He saw the disc of Jupiter with its satellites, and

The reflecting telescope.

Venus distinctly horned, like a crescent moon. Then he lent it to Barrow. Barrow carried it to London, to show his friends at the Royal Society.

Like no institution before it, the Royal Society was born dedicated to information flow. It exalted communication and condemned secrecy. 'So far are the narrow conceptions of a few private Writers, in a dark Age, from being equal to so vast a design,' its founders declared. Science did not exist – not as an institution, not as an activity – but they conceived it as a public enterprise. They imagined a global network, an 'Empire in Learning'. Those striving to grasp the whole fabric of nature

ought to have their eyes in all parts, and to receive information from every quarter of the earth, they ought to have a constant universal intelligence: all discoveries should be brought to them: the Treasuries of all former times should be laid open before them.[18]

And in what language? The society's work included translation, contending with scores of vernacular dialects in Europe, and even stranger languages were reported to exist in faraway India and Japan. Latin served for standardisation, but the society's founders explicitly worried about the uses of any language. Philosophy had mired itself in its own florid eloquence. They sought 'not the Artifice of Words, but a bare knowledge of things'. Now it was time for plain speaking, the most naked expression, and when possible this meant the language of mathematics.[19]

Words were truant things, elusive of authorities, malleable and relative. Philosophers had much work to do merely defining their terms, and words like *think* and *exist* and *word* posed greater challenges than *tree* and *moon*. Thomas Hobbes warned:

The light of humane minds is perspicuous words, but by exact definitions first snuffed, and purged from ambiguity; reason is the pace . . . And, on the contrary, metaphors, and senseless and ambiguous words are like *ignes fatui*; and reasoning upon them is wandering amongst innumerable absurdities.[20]

Galileo, having observed sunspots through his telescope in 1611, could not report the fact without entering a semantic thicket:

So long as men were in fact obliged to call the sun 'most pure and most lucid', no shadows or impurities whatever had been perceived in it; but now that it shows itself to us as partly impure and spotty; why should we not call it 'spotted and not pure'? For names and attributes must be accommodated to the essence of things, and not the essence to the names, since things come first and names afterwards.[21]

It has always been so – this is the nature of language – but it has not always been equally so. Diction, grammar and orthography were fluid; they had barely begun to crystallise. Even proper names lacked approved spelling. Weights and measures were a hodgepodge. Travellers and mail made their way without *addresses*, unique names and numbers as coordinates for places. When Newton sent a letter to the Secretary of the Royal Society, he directed it *To Mr Henry Oldenburge at his house about the middle of the old Palmail in St Jamses Fields in Westminster.*[22]

Oldenburg was an apostle for the cause of collective awareness – born Heinrich Oldenburg in the trading city of Bremen (he was never sure what year), later Henricus, and now Henry. He had come to England during the Civil War as an envoy on a mission to Oliver Cromwell. He began corresponding with learned men such as Cromwell's Latin Secretary, John Milton; Cromwell's brother-in-law, John Wilkins; the young philosopher Robert Boyle; and others – soon to be the nucleus of the Royal Society. Then, as an acquaintance put it, 'this Curious German having well improved himself by his Travels, and... rubbed his Brains against those of other People, was... entertained as a Person of great Merit, and so made Secretary to the Royal Society.'[23] He was a master of languages and the perfect focal point for the society's correspondence. He employed both the ordinary post and a network of diplomatic couriers to receive letters from distant capitals, especially Paris and Amsterdam. In 1665 he began printing and distributing this correspondence in the form of a news sheet, which he called the *Philosophical Transactions.* This new creature, a journal of science, remained Oldenburg's personal enterprise till the end of his life.[24] He found a printer and stationer with carriers who could distribute a few hundred copies across London and even further.

The news took many forms. Mr Samuel Colepress, near Plymouth, reported his observations of the height and velocity of the daily tides; from March to September, he asserted, the tides tended to be a foot higher ('*perpendicular*, which is always to be understood') in the morning than in the evening.[25] An author in Padua, Italy, claimed to have discovered new arguments against the motion of the earth, and a mathematician there disputed him, citing an experiment by a Swedish gentleman, who fired shots from 'a Canon perpendicular to the Horizon' and observed whether the balls fell towards the west or the east. Mr Hooke saw a spot on the planet Jupiter. A very odd monstrous calf was born in Hampshire. A newly invented instrument of music arrived: a harpsichord, with gut-strings. There were poisonous vipers and drops of poison from Florence. The society examined the weaving of asbestos – a cloth said to endure the fiercest fire – and models of perpetual motion.[26]

No sooner had the virtuosi begun to gather than England's poets satirised their fixations and their questions. Hooke himself made an easy target – his fantastic world of fleas and animalcules. The natural philosopher could easily be portrayed as a preoccupied pedant, and not so easily distinguished from the astrologer and the alchemist. 'Which way the dreadful comet went / In sixty-four and what it meant?' asked Samuel Butler (his mockery tinged with wonder).

> Whether the Moon be sea or land
> Or charcoal, or a quench'd firebrand . . .
> These were their learned speculations
> And all their constant occupations,
> To measure wind, and weigh the air
> And turn a circle to a square.[27]

In fact, travel and trade, more than speculation or technology, fuelled the society's business; bits of exotic knowledge came as fellow travellers on ships bearing foreign goods. Spider webs were seen in faraway Bermuda and 300-foot cabbage trees in the Caribe Islands.[28] A worthy and inquisitive gentleman, Captain Silas Taylor of Virginia, reported that the scent of the wild Penny-royal could kill Ratle-Snakes. A German Jesuit, Athanasius Kircher, revealed secrets of the subterranean world: for example, that the ocean waters continually pour into the northern pole, run through the bowels of the earth, and regurgitate at the southern pole.

Far away in Cambridge Newton inhaled all this philosophical news. He took fervid notes. Rumours of a fiery mountain: 'Batavia one afternone was covered with a black dust heavyer then gold which is thought came from an hill on Java Major supposed to burne.'[29] Rumours of lunar influence: 'Oysters & Crabs are fat at the new moone & leane at the full.' Then in 1671 he heard directly from the voice of the Royal Society. 'Sr,' Oldenburg wrote, 'Your Ingenuity is the occasion of this addresse by a hand unknowne to you . . .'

He said he wished to publish an account of Newton's reflecting telescope. He urged Newton to take public credit. This peculiar historical moment – the manners of scientific publication just being born – was alert to the possibilities of plagiarism. Oldenburg raised the spectre of ' the usurpation of foreigners' who might already have seen Newton's instrument in Cambridge, 'it being too frequent, the new Inventions and contrivances are snatched away from their true Authors by pretending bystanders'.[30] The philosophers were proposing Newton for election as a fellow of the society. Still, there were questions. Some of the skilful examiners agreed that Newton's tube magnified more than larger telescopes, but others said this was hard to measure with certainty.[31] Some, ill at ease with the technology, complained that such a powerful telescope made it difficult 'to find the Object'. Meanwhile Hooke told the members privately that he himself had earlier made a much more powerful tiny telescope, in 1664, just an inch long, but that he had not bothered to pursue it because of the plague and the fire. Oldenburg chose not to mention Hooke's claim.

Newton wrote back with conventional false modesty:

I was surprised to see so much care taken about securing an invention to mee, of which I have hitherto had so little value. And therefore since the R. Society is pleased to think it worth the patronizing, I must acknowledg it deserves much more of them for that, then of mee, who, had not the communication of it been desired, might have let it still remained in private as it hath already done some yeares.[32]

A fortnight later he set modesty aside. He wished to attend a meeting, he told Oldenburg dramatically.

I am purposing them, to be considered of & examined, an accompt of a Philosophicall discovery which induced me to the making of the said Telescope, & which I doubt not but will prove much more gratefull then the communication of that instrument, being in my Judgment the oddest if not the most considerable detection which hath hitherto been made in the operations of Nature.[33]

And by the way, what would his duties be, as Fellow of the Royal Society?

7

RELUCTANCY AND
REACTION

The great court of Trinity College was mostly complete, with a library and stables, central fountain, and fenced-in plots of grass. An avenue of newly planted linden trees lay to the southwest.[1] Newton occupied a chamber upstairs between the Great Gate and the chapel. To the west stood a four-walled court used for the game of tennis. Sometimes he watched fellows play, and he noticed that the ball could curve, and not just downward. He understood intuitively why this should be so: the ball was struck obliquely and acquired spin. 'Its parts on that side, where the motions conspire, must press and beat the contiguous Air more violently than on the other, and there excite a reluctancy and reaction of the Air proportionately greater.'[2] He noted this in passing because he had wondered whether rays of light could swerve the same way – if they 'should possibly be globular bodies' spinning against the ether. But he had decided against that possibility.

He did not go to London to appear before the Royal Society after all – not for three more years – but he did not wait to send Oldenburg his promised account of a philosophical discovery. He composed a long letter in February 1672, to be read aloud at a meeting. Within a fortnight Oldenburg had it set in type and printed in the *Philosophical Transactions*, along with a description of the East Indian coasts and an essay on music.[3]

Newton's letter presented both an experiment and a 'theory'.[4] Six years before, he wrote, he had aligned his prism in a sunbeam entering a dark room through a hole in the window shutter. He expected to see all the colours of the rainbow fanned against the wall and, indeed, he did – vivid and intense, a very pleasing divertissement, he reported. This phenomenon of colours was ancient. As soon as people had glass – that is, as soon as they had *broken* glass – they noticed

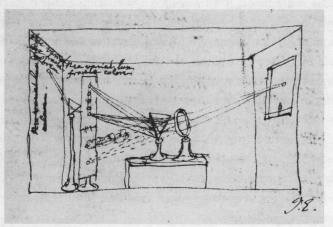

Experimentum Crucis: The sunbeam from the window shutter passes through one prism, separating it into colours; then a beam of coloured light passes through a second prism. The second prism has no further separation to perform: the white light is a mixture, but the coloured beams are pure.

the appearance of colours where two refracting surfaces formed a sharp edge.[5] A carefully formed triangular prism manifested colours most perfectly. No one knew where the colours came from, but it had seemed clear enough, almost by definition, that a prism somehow *creates* colours.

Newton noted a surprise (or so he claimed): where he would have expected the refracted light to form a circle on the wall – all the sun's rays being refracted equally – instead he saw an oblong. He tried moving the prism, to see whether the thickness of the glass made a difference. He tried varying the size of the hole in the window shutter. He tried a second prism. He measured the distance from the aperture to the wall (22 feet); the length of the coloured oblong (13¼ inches); its width (2⅝ inches); and the angles of incidence and refraction, known to be mathematically linked. He noted that the sun was not a point but a disc, spread across 31 minutes of arc. The sunbeam was always in motion, and he could examine it only for moments at a time, but he did not let go of this small oddity – this peculiar elongation of the image.

It led him (or so he reported) to the *Experimentum Crucis* – the signpost at a crossroads, the piece of experience that shows which path to trust. Newton took the high-plumed phrase from Hooke, who had adapted it from Bacon.[6] The crucial idea was to isolate a beam of *coloured* light and send that through a prism. For this he needed a pair of prisms and a pair of boards pierced with holes. He aligned these and carefully rotated one prism in his hand, directing first blue light and then red light through the second prism. He measured the angles: the blue rays, bent slightly more by the first prism, were again refracted slightly more by the second. Most persuasive, though, was that the second prism never created new colours or altered the colours shining from the first prism. Years before, in his earliest speculation, he had asked himself, 'Try if two Prismas the one casting blue upon the other's red doe not produce a white.'' They did not. Blue light stayed blue and red stayed red. Unlike white (Newton deduced) those colours were pure.

'And so the true cause of the length of that Image was detected,' Newton declared triumphantly – 'that *Light* consists of *Rays differently refrangible*.' Some colours are refracted more, and not by any quality of the glass but by their own predisposition. Colour is not a modification of light but an original, fundamental property.

Above all: white light is a heterogeneous mixture.[8]

But the most surprising, and wonderful composition was that of *Whiteness*. There is no one sort of Rays which alone can exhibit this. 'Tis ever compounded, and to its composition are requisite all the aforesaid primary Colours, mixed in due proportion. I have often with Admiration beheld, that all the Colours of the Prisme being made to converge, and thereby to be again mixed, . . . reproduced light, intirely and perfectly white.

A prism does not create colours; it separates them. It takes advantage of their different refrangibility to sort them out.

Newton's letter was itself an experiment, his first communication of scientific results in a form intended for publication.[9] It was meant to persuade. He had no template for such communication, so he invented one: an autobiographical narrative,

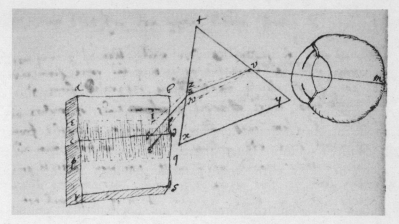

A prism refracts blue light more than red.

step by step, actions wedded to a sequence of reasoning. He exposed intimate feelings: his pleasure at the display of colours, his uncertainty, and then above all his surprise and wonder.

The account was an artifice, stylising a process of discovery actually carried out over years, on odd occasions, sometimes below the level of consciousness and computation. A prism in a pencil-thin sunbeam actually makes a smudge of colour on a wall, uneven and unstable, its edges shadowy and fading. He idealised what he described; the image made sense only because he already knew what he was looking for. He had already seen, years before, that blue light is bent more than red; he had looked through a prism at red and blue threads and noted their varying refraction. He also knew that refracting lenses smeared colours; that was why he had invented a reflecting telescope.

When Descartes looked at a prism in sunlight, he had seen a circle of colours, not an oblong. A circle was the shape he expected, and it was tiny, because he directed his prism's light at nearby paper, not a wall twenty-two feet distant. Newton wanted to see the oblong, the spreading; he wanted to magnify it; he wanted to measure it against his geometrical intuition about the laws of refraction; he believed in precision and in his ability to interpret small discrepancies. Indeed, he believed in mathematics as the road to understanding, and he said so: that he expected even the science of colours to become mathematical. And this meant *certain.* 'For what I shall tell concerning them is not an Hypothesis but most rigid consequence,' he wrote, 'not conjectured by barely inferring tis thus because not otherwise . . . but evinced by the mediation of experiments concluding directly & without suspicion of doubt.'[10] Oldenburg omitted this sentence from the version he printed.

What was light, anyway? In this offering of a 'theory', Newton chose not quite to commit himself, but he had a mental picture: a ray of light was a stream of particles, 'corpuscles' – material substance in motion. Descartes had thought light was pressure in the ether and colour an effect of the rotation of these ether particles; Hooke objected to that and proposed the notion of light as a pulse, a vibration of the ether, or a wave, like sound. Newton found Hooke's theory galling. 'Though Descartes may bee mistaken so is Mr Hook,' he wrote privately, in taking notes on his copy of *Micrographia.* He had a simple argument against a wave theory: light (unlike sound) does not turn corners. 'Why then may not light deflect from straight lines as well as sounds &c?'[11] In his notes Newton wrote of light as globules, travelling at finite speed and impinging on the eye. In his letter he stuck abstractly to rays. 'To determine more absolutely, what Light is, . . . and by what modes or actions it produceth in our minds the Phantasms of Colours, is not so easie. And I shall not mingle conjectures with certainties.'[12]

Certainties or not, Newton's conclusions represented a radical assault on the prevailing wisdom.[13] For the next four years the *Philosophical Transactions* boiled with controversy, month after month: ten critiques of Newton's letter and eleven counters from Newton.[14] Oldenburg kept assuring him of the society's applause for his ingenuity and frankness and its concern that the honour of discovery might be snatched from him and assumed by foreigners.[15] In his role as a clearing house for developments in mathematics, Oldenburg discovered that he could use discoveries by foreigners – for example, Gottfried Wilhelm Leibniz in Germany – to pry secret knowledge from Newton. He grew used to Newton's tantalising style, always holding gems just out of reach.

And in fact I know myself how to form a series . . .

I cannot proceed with the explanation of it now . . .

I have preferred to conceal it thus . . .

Once this was known, that other could not long remain hidden from me . . .

I have another method not yet communicated, . . . a convenient, rapid and general solution of this problem, *To draw a geometrical curve which shall pass through any number of given points* . . . These things are done at once geometrically with no calculation intervening . . . Though at first glance it looks unmanageable, yet the matter turns out otherwise. For it ranks among the most beautiful of all that I could wish to solve.[16]

His mathematics remained mostly hidden. Regarding light, however, he had exposed himself, and he regretted it. Hooke continued to attack. As Curator of Experiments Hooke assured the society that he had already performed these very experiments, hundreds of times. He was not a little pleased, he said, with the niceness and curiosity of Newton's observations, but he had to confess that he considered these arguments a mere hypothesis. He said that his own experiments – 'nay and even those very expts which he alledged' – proved that light is a pulse in the ether and that colour is nothing but a disturbance of that light. He would be glad to see 'one Experimentum crucis from Mr Newton' to make him change his mind, but this was not it. A prism adds colour to light, he insisted, just as an organ pipe or a violin string adds sound to the air.[17] A French Jesuit, Ignace Pardies, wrote from Paris that Newton's 'hypothesis' would overthrow the very basis of optics; that the oblong image could be explained by rays coming from different parts of the sun's face; and that mixing coloured rays of light produces only a dark blur, not white.[18]

All this angered Newton, especially the word *hypothesis*. He was not offering a hypothesis, he said again, but 'nothing else than certain properties of light which, now discovered, I think are not difficult to prove, and which if I did not know to be true, I should prefer to reject as vain and empty speculation, than acknowledge them as my hypothesis.'[19] Oldenburg suggested that he respond without mentioning names – especially Hooke's – but Newton had a different idea. Months went by, and his rancour festered. When he finally penned a long reply, it named Hooke in its first sentence and on every page. 'I was a little troubled to find a person so much concerned for an *Hypothesis*,' he wrote, 'from whome in particular I most expected an unconcerned & indifferent examination.'

Mr Hook thinks himselfe concerned to reprehend me . . . But he knows well that it is not for one man to prescribe Rules to the studies of another, especially not without understanding the grounds on which he proceeds. Had he obliged me with a private letter . . .[20]

Hooke's rejection of the *experimentum crucis* was 'a bare denyall without assigning a reason', he asserted. Newton wrote and rewrote this letter four times. It grew far longer than his original report. He considered colours in bubbles and froth; jabbed slyly at Hooke with suggestions for microscopy; and refined his distinction between pure colours and compounded whiteness. There were many ways to mix colours, he suggested, to produce white or (not so perfect and intense) grey. 'The same may be effected by painting a *Top* (such as Boys play with) of divers colours, for when it is made to circulate by whipping it will appear of such a dirty color.'

Above all, he wished to assert that optics was a mathematical science, rigorous and certain; that it depended on physical principles and mathematical proof; and that since he had learned these principles he had met with constant success.

He implied again and again that Hooke was not really performing the experiments. Hooke had 'maimed' his argument. Hooke insisted on 'denying some things the truth of which would have appeared by an experimentall examination'. True – Newton conceded – he was arguing for the corporeity of light, but that followed from his theory, not the other way around. It was not a fundamental supposition. In suggesting that light was composed of particles, he had carefully used the word *perhaps*. 'I wonder how Mr Hook could imagin that when I had asserted the Theory with the greatest rigor, I should be so forgetfull as afterwards to assert the fundamentall supposition it selfe with no more than a *perhaps*.'

Hooke was Newton's most enthusiastic antagonist now, but not his most able. Christiaan Huygens, the great Dutch mathematician and astronomer, also favoured a wave theory of light. His understanding of refraction and reflection was profound – and correct enough, when alloyed with Newton's, to survive up to the quantum era. But he, too, by way of letters to Oldenburg, raised initial questions about Newton's 'hypothesis' and in return felt the young man's wrath. He caught subtle errors that Newton would never quite acknowledge; for example, Huygens suggested correctly that white could be created not just by a mixture of all colours but by the blending of pairs such as blue and yellow.[21] Fifteen months after his election to the Royal Society, Newton announced that he wished to withdraw – and not just from the society but from all correspondence. 'I suppose there hath been done me no unkindness,' he wrote Collins. 'But I could wish I had met with no rudeness in some other things. And therefore I hope you will not think it strange if to prevent accidents of that nature for the future I decline that conversation which hath occasioned what is past.'

Oldenburg begged him to reconsider, suggested he no longer feel obliged to pay his dues, and assured him that the Royal Society esteemed and loved him.[22] The criticism had been so mild and so ordinary, though perhaps there had been 'incongruities'. Newton had still never met any of these men – Oldenburg, Collins, Hooke, or the others. He wrote one more reply. 'The incongruities you speak of, I pass by,' he said. 'But . . . I intend to be no further sollicitous about matters of Philosophy. And therefore I hope you will not take it ill if you find me ever refusing doing any thing more in that kind.'[23] Oldenburg did not hear from him again for more than two years.[24]

He had discovered a great truth of nature. He had proved it and been disputed. He had tried to show how science is grounded in concrete practice rather than grand theories. In chasing a shadow, he felt, he had sacrificed his tranquillity.[25]

8

IN THE MIDST OF
A WHIRLWIND

When he observed the world it was as if he had an extra sense organ for peering into the frame or skeleton or wheels hidden beneath the surface of things. He sensed the understructure. His sight was enhanced, that is, by the geometry and calculus he had internalised. He made associations between seemingly disparate physical phenomena and across vast differences in scale. When he saw a tennis ball veer across the court at Cambridge, he also glimpsed invisible eddies in the air and linked them to eddies he had watched as a child in the rock-filled stream at Woolsthorpe. When one day he observed an air pump at Christ's College, creating a near vacuum in a jar of glass, he also saw what could not be seen, an invisible negative: that the reflection on the inside of the glass did not appear to change in any way. No one's eyes are that sharp. Lonely and dissocial as his world was, it was not altogether uninhabited; he communed night and day with forms, forces and spirits, some real and some imagined.

In 1675 Newton journeyed to London and finally appeared at the Royal Society. He met in person these men who had till then been friends and antagonists twice removed, their spirits channelled through Oldenburg's mail. Among the virtuosi made flesh was Robert Boyle, fifteen years his senior and a mentor of Hooke's. Boyle was a fervent corpuscularian; in his great polemic *The Sceptical Chymist* he had developed a theory of fundamental particles as the constituents of matter. He believed that all the phenomena of nature could be explained by the combination and organisation of these atoms into mixed bodies, some perfect and some imperfect and none more perfect than gold.[1] He believed in the alchemists' greatest dream, the transmutation of baser metals into gold, but he reviled their traditions of secrecy – 'their obscure, ambiguous, and almost Ænigmatical Way of expressing what they pretend to Teach'.[2]

They have no mind to be understood at all, but by the Sons of Art (as they call them) nor to be understood even by these without Difficulty and Hazardous Tryalls.

His experiments with an air pump were renowned, and his own investigation of colour had spurred Hooke and Newton in turn. He greeted Newton warmly.

Over the next months Newton, back in Cambridge, laboured over a new manuscript. He set down in passionate words his own corpuscular theory. Here, finally, was his *Hypothesis* – he embraced the label he had denied so vehemently before. 'An Hypothesis,' he titled it, 'explaining the Properties of Light discoursed of in my severall Papers'.[3] But he spoke of more than light alone; he was taking on the whole substance of nature. His nemesis, Hooke, loomed large. 'I have observed the heads of some great virtuoso's to run much upon Hypotheses,' Newton said, 'as if my discourses wanted an Hypothesis to explain them by.' He noted that 'some' could not quite take his meaning when he spoke of light and colour in the abstract, and perhaps they would understand better with an illustration. Thus – the 'Hypothesis'.

He wanted Oldenburg to read this to the assembled Royal Society but not to publish it. And he wanted his listeners to understand a delicate rhetorical point. He did not pretend to mathematical certainty here, even if, for convenience,[4] he chose to 'speak of it as if I assumed it & propounded it to be beleived'. Let no man 'think me oblig'd to answer objections against this script,' he said. 'For I desire to decline being involved in such troublesome & insignificant Disputes.'

This sheaf of papers posted to Oldenburg[5] blended calculation and faith. It was a work of the imagination. It sought to reveal nothing less than the microstructure of matter. For generations it reached no further than the few men who heard it read and then raptly debated it through all the meetings of the Royal Society from December 1675 to the next February. Newton had peered deeper into the core of matter than could be justified by the power of microscopes. Through a series of experiments and associations he seemed to feel nature's fundamental particles just beyond the edge of his vision. Indeed, he predicted that instruments magnifying three or four thousand times might bring atoms into view.[6]

He saw a vast range of phenomena to explain, and the cool certainties of geometry had reached the limit of their usefulness here. There were all kinds of chemical activity, processes like vegetation, fluids that interacted with more or less 'sociableness'. He closed his eyes to no problem because it was too mysterious or intractable. He confounded the distant members of the society with a vivid description of an experiment revealing *electricity*, a power certain bodies gained when excited: he rubbed a glass disc with cloth and then waved it over bits of paper. They sprang to life:

Sometimes leaping up to the Glass & resting there a while, then leaping downe & resting there, then leaping up & perhaps downe & up again ... sometimes in lines perpendicular to the Table, Sometimes in oblique ones ... & turn often about very nimbly as if ... in the midst of a whirlwind.[7]

Irregular motions, he emphasised – and he saw no way to explain them mechanically, purely in terms of matter pressing on matter. It was no static world, no orderly world he sought to understand now. Too much to explain at once: a world in flux; a world of change and even chaos. He gave out poetry:

For nature is a perpetuall circulatory worker, generating fluids out of solids, and solids out of fluids, fixed things out of volatile, & volatile out of fixed, subtile out of gross, & gross out of subtile, Some things to ascend & make the upper terrestriall juices, Rivers and the Atmosphere; and by consequence others to descend ...[8]

The ancients had often supposed the existence of ether, a substance beyond the elements, purer than air or fire. Newton offered the ether as a hypothesis now, describing it as a 'Medium much of the same constitution with the air, but far rarer, subtiler & more strongly Elastic'. As sound is a vibration of the air, perhaps there are vibrations of the ether – these would be swifter and finer. He estimated the scale of sound waves at a foot or half-foot, vibrations of ether at less than a hundred thousandth of an inch.

This ether was a philosophical hedge, a way of salvaging a mechanical style of explanation for processes that seemed not altogether mechanical: iron filings near a magnet arrange themselves into curved lines, revealing 'magnetic effluvia'; chemical change occurs in metals even after they have been sealed in glass; a pendulum swings far longer in a glass emptied of air, but ceases eventually nonetheless, proving that 'there remains in the glass something much more subtle which damps the motion of the bob'.[9] The mechanists were labouring to banish occult influences – mysterious action without contact. The ether, more subtle than air, yet still substantial, might convey forces and spirits, vapours and exhalations and condensations. Perhaps an ethereal wind blew those fluttering bits of paper. Perhaps the brain and nerve transmitted ethereal spirit – the soul inspiring muscle by impelling it through the nerves.[10] Perhaps fire and smoke and putrefaction and animal motion stemmed from the ether's excitation and swelling and shrinking. Perhaps this ether served as the sun's fuel; the sun might imbibe the ethereal spirit 'to conserve his Shining, & keep the Planets from recedeing further from him'.[11] (The apple had dropped long since, but universal gravitation remained remote.)

Hooke, listening to Oldenburg read Newton's words aloud, kept hearing his name. 'Mr Hook, you may remember, was speaking of an odd straying of light ... near the edge of a Rasor ...' Indeed, earlier in 1675 Hooke had put forward his new discovery of the phenomenon later known as diffraction: the bending of light at a sharp edge. One way to explain diffraction – the only way, until quantum mechanics – was in terms of the interference of waves. Did this spreading of light rays mean that they could curve after all, as sound waves apparently do around corners? Newton said he was unsure: 'I took it to be onely a new kind of refraction, caused perhaps by the externall æthers beginning to grow rarer a little before it come at the Opake body ...' He recalled, though, that Hooke had been

pleased to answer that though it should be but a new kind of refraction, yet it was a *new one*. What to make of this unexpected reply, I knew not, haveing no other thoughts but that a new kind of refraction might be as noble an Invention as any thing els about light.

A noble invention, Newton agreed. But he remembered having read about this experiment before Hooke's account. He was obliged to mention that the French Jesuit Honoré Fabri had described it; and Fabri in turn had got it from a Bolognese mathematician, Francesco Maria Grimaldi.[12] It was not Hooke's discovery.

Hooke grew irate. In evenings that followed he met with friends in coffee houses and told them that Newton had commandeered his pulse theory. After all, Newton was talking about colour in terms of 'vibrations of unequal bignesses'. Large vibrations are red – or, as he said more carefully, cause the sensation of red. Short vibrations produce violet. The only difference between colours was this: a slight, quantifiable divergence in the magnitude of vibration. Newton did not speak of *waves*. Nor for that matter had Hooke: waves were still a phenomenon of the sea. A lack of vocabulary hindered both men; but what Newton had seen was just what Hooke had sought.

This was insupportable. At the end of the second meeting devoted to the Newton 'Hypothesis', Hooke rose to declare that the bulk of it had come from his *Micrographia*, 'which Mr Newton had only carried farther in some particulars'.[13] Oldenburg lost no time in reporting this claim back to Cambridge.

Cambridge fired back. 'As for Mr Hook's insinuation,' Newton wrote Oldenburg, 'I need not be much concerned at the liberty he takes.'[14] He wished to avoid 'the savour of having done any thing unjustifiable or unhansome towards Mr Hook'. So he analysed the chain of logic and priority. First, what is actually Hooke's? We must 'cast out what he has borrowed from Des Cartes or others':

That there is an ether. That light is the action of this ether. That the ether penetrates solid bodies in varying degrees. That light is at first uniform. That colours come from a modification of light rays – accelerated to make red and retarded to make blue, all other colours coming from some mixture of red and blue.

All Hooke did was change Descartes's idea of a pressing motion in the ether to a vibrating one. Globules for Descartes, pulses for Hooke. 'In all this,' Newton concluded,

I have nothing common with him but the supposition that æther is a Medium susceptible of vibrations of which supposition I make a very different use: he supposing it light it self which I suppose it not.

For the rest – refraction and reflection and the production of colours – Newton said he explained it all so differently from Hooke as to 'destroy all he has said'. He added sarcastically, 'I suppose he will allow me to make use of what I tooke the pains to find out.'

Hooke was poking at a soft spot in Newton's understanding of light. Was it particle or wave? Newton was vacillating on this matter now, as humanity would continue to vacillate until twentieth-century physicists vanquished the paradox by accepting it. Newton both exposed his uncertainty and concealed it. He played a delicate game, ringing changes on the word *hypothesis*, trying to distinguish between what he knew and what he was forced to suppose. He supposed the existence of an ether – mysterious and even spiritual – because he could not dispense with such a thing, for now.

Oldenburg – no friend to Hooke[15] – chose to surprise him with a public reading of Newton's rejoinder at the next Royal Society meeting. Finally, after years of jousting by proxy, Hooke decided to take pen in hand and address his adversary personally.[16] He adopted a meek and philosophical tone. He said he suspected Newton was being misinformed; he had experience with such 'sinister practices'. He did not wish to contend or feud or be 'drawn to such kind of warr'. We are 'two hard-to-yield contenders', he proclaimed. 'Your Designes and myne I suppose aim both at the same thing which is the Discovery of truth and I suppose we can both endure to hear objections.'

Newton's famous reply came a fortnight later.[17] If the weapons of this duel were to be insincere politesse and exaggerated deference, he could wield them as well. He called Hooke a 'true Philosophical spirit'. He gladly embraced the proposal of a private correspondence. 'What's done before many witnesses is seldome without some further concern than that for truth: but what passes between friends in private usually deserves the name of consultation rather than contest, & so I hope it will prove between you & me.' And then, for the matter of their dispute, he put on record a finely calibrated piece of faint praise and lofty sentiment:

What Des-Cartes did was a good step. You have added much several ways, & especially in taking the colours of thin plates into philosophical consideration. If I have seen further it is by standing on the sholders of Giants.[18]

The private philosophical dialogue between Newton and Hooke never took place. Almost two years passed before they communicated again at all. By then Oldenburg had died, Hooke had succeeded him as Secretary of the Royal Society, and Newton had withdrawn ever more deeply into the seclusion of his Trinity chambers.

9

ALL THINGS ARE
CORRUPTIBLE

His devotion to philosophical matters grew nonetheless. He built a special chimney to carry away the smoke and fumes.[1]

By Newton's thirties his hair was already grey, falling to his shoulders and usually uncombed. He was thin and equine, with a strong nose and gibbous eyes. He stayed in his chamber for days at a time, careless of meals, working by candlelight. He was scarcely less isolated when he dined in the hall. The fellows of Trinity College learned to leave him undisturbed at table and to step around diagrams he scratched with his stick in the gravel of the walkways.[2] They saw him silent and alienated, with shoes down at heel and stockings untied. He feared disease – plague and pox – and treated himself pre-emptively by drinking a self-made elixir of turpentine, rosewater, olive oil, beeswax and sack. In fact he was poisoning himself, slowly, by handling mercury.[3]

No one could understand till centuries later – after his papers, long hidden and scattered, began finally to be reassembled – that he had been not only a secret alchemist but, in the breadth of his knowledge and his experimentation, the peerless alchemist of Europe. Much later, when the age of reason grew mature, a fork was seen to have divided the road to the knowledge of substances. On one path, chemistry: a science that analysed the elements of matter with logic and rigour. Left behind, alchemy: a science and an art, embracing the relation of the human to the cosmos; invoking transmutation and fermentation and procreation. Alchemists lived in a realm of exuberant, animated forces. In the Newtonian world of formal, institutionalised science, it became disreputable.

But Newton belonged to the pre-Newtonian world. Alchemy was in its heyday. A squalid flavour did attach to such researches; alchemists were suspected as charlatans pretending to know how to make gold. Yet the modern distinction between chemistry and alchemy had not emerged. When the vicar John Gaule, an expert on witchcraft, assailed 'a kinde of præstigious, covetous, cheating magick', he called this malodorous practice by its name: *chymistry*.[4] If alchemists were known to treasure secrecy and obscure their writings with ciphers and anagrams, these habits were no bar for Newton, burrowing further inward. If they revered arcane authorities and certain sacred texts, if they adopted Latinate pseudonyms and circulated secret manuscripts, so for that matter did Christian theologians. Newton was a mechanist and a mathematician to his core, but he could not believe in a nature without spirit. A purely mechanical theory for the world's profusion of elements and textures – and for their transformations, from one substance to another – lay too far beyond reach.

He met with mysterious men and copied their papers – a W.S., a Mr. F.[5] He devised a pseudonym, *Jeova sanctus unus*, an anagram of Isaacus Neuutonus. In the garden outside his room he built a laboratory, a shed abutting the wall of the chapel. His fire burned night and day.[6] To alchemists nature was alive with process. Matter was active, not passive; vital, not inert. Many processes began in the fire: melting, distilling, subliming and calcining. Newton studied them and practised them, in his furnaces of tin and bricks and firestones. In sublimation vapours rose from the ashes of burned earths and condensed again upon cooling. In calcination fire converted solids to dust; 'be you not weary of calcination', the alchemical fathers had advised; 'calcination is the treasure of a thing'.[7] When a crimson-tinged earth, cinnabar, passed through the fire, a coveted substance emerged: 'silvery water' or 'chaotic water' – quicksilver.[8] It was a liquid and a metal at once, lustrous white, eager to form globules. Some thought a wheel rimmed with quicksilver could turn unaided – perpetual motion.[9] Alchemists knew quicksilver as Mercury (as iron was Mars, copper Venus, and gold the sun); in their clandestine writings they employed the planet's ancient symbol, $\mathemph{☿}$. Or they alluded to quicksilver as 'the serpents'.[10]

'The two serpents ferment well . . .' Newton wrote at one session. 'When the fermentation was over I added $☿$ 16^{gr} & the matter swelled much with a vehement fermentation . . .'[11] Like other alchemists, he conceived of mercury not just as an element but as a state or principle inherent in every metal. He spoke of the 'mercury' of gold. He particularly coveted a special, noble, 'philosophical' mercury: 'this $☿$ . . . drawn out of bodies hath as many cold superfluities as common $☿$ hath, & also a special form & qualities of the metals from which it was extracted'.[12] Part of mercury's esoteric appeal was its tendency to react with other metals. Applied to copper, lead, silver and even gold, it formed soft amalgams. A skilful practitioner could use mercury to purify metals. Over time, mercury builds up in the body, causing neurological damage: tremors, sleeplessness, and sometimes paranoid delusions.

Robert Boyle, too, was experimenting with mercury. In the spring of 1676, Newton read in the *Philosophical Transactions* an account 'Of the Incalescence of Quicksilver with Gold, generously imparted by B.R.'[13] He recognised the inverted initials, and he suspected that the research drew near the alchemists' dream of multiplying gold. 'I believe the fingers of many will itch to be at the knowledge of the preparation of such a ☿,' he wrote privately. A dangerous sort of knowledge might lie nearby – 'an inlet to something more noble, not to be communicated without immense damage to the world'.[14] Newton believed – and knew Boyle did, too – that the basic substance of matter was everywhere the same; that countless shapes and forms flowed from the varied operations of nature on this universal stuff. Why should the transmutation of metals be impossible then? The history of change was all around.

Like no other experimenter of his time, alchemist or chemist, he weighed his chemicals precisely, in a balance scale.[15] Obsessed as always with the finest degrees of measurement, he recorded weights to the nearest quarter of a grain. He measured time, too; here, a precise unit was an eighth of an hour. But measurement never replaced sensation: as his experiments fumed, he touched and sniffed and tasted the slimes and liquors that emerged.

He probed for the processes of life and death: vegetation and, a special case, putrefaction, which produces a 'blackish rotten fat substance' and exhales matter into fumes. *Nothing can be changed from what it is without putrefaction,* he wrote in haste, in his microscopic scrawl. First nature putrefies, then it generates new things. *All things are corruptible. All things are generable.* And so the world continually dies and is reborn. These exhalations, and mineral spirits, and watery vapours, generate a rising air and buoy up the clouds: 'so high as to loos their gravity'.[16]

This is very agreeable to natures proceedings to make a circulation of all things. Thus this Earth resembles a great animall or rather inanimate vegetable, draws in æthereall breath for its dayly refreshment and vital ferment ... This is the subtil spirit which searches the most hiden recesses of all grosser matter which enters their smallest pores and divides them more subtly then any other materiall power what ever.

Driving this cycle of death and life, inspiring this circulatory world, must be some active spirit – nature's universal agent, her secret fire. Newton could not but identify this spirit with light itself – and light, in turn, with God. He marshalled reasons. All things, in the fire, can be made to give off light. Light and heat share a mutual dependence. No substance so subtly pervades all things as light. He felt this in the depth of his being.

'Noe heat is so pleasant & beamish as the suns,' he wrote.

Through his alchemical study shines a vision of nature as life, not machine. Sexuality suffused the language of alchemy.

Generation came from seed and copulation; principles were male (Mercury) and female (Venus). Then again:

these two mercuries are the masculine and feminine semens ... fixed and volatile, the Serpents around the Caduceus, the Dragons of Flammel. Nothing is produced from masculine or feminine semen alone ... The two must be joined.[17]

From the seeds, the seminal virtues, came the fire and the soul. If alchemy was the closest Newton came to a worldly exploration of sexuality, it crossed paths with a theological quest as well. To alchemists the transmutation of metals meant a spiritual purification. It was God who breathed life into matter and inspired its many textures and processes. Theology joined alchemy as the chief preoccupation of Newton's middle decades.

The new mechanical philosophers, striving to create a science free of occult qualities, believed in matter without magic – inanimate brute matter, as Newton often called it. The virtuosi of the Royal Society wished to remove themselves from charlatans, to build all explanations from reason and not miracles. But magic persisted. Astronomers still doubled as astrologers; Kepler and Galileo had trafficked in horoscopes.[18] The magician, probing nature's secrets, served as a template for the scientist. 'Do you believe then,' Nietzsche asked two centuries later, 'that the sciences would ever have arisen and become great if there had not beforehand been magicians, alchemists, astrologers and wizards, who thirsted and hungered after abscondite and forbidden powers?'

Descartes had gone to great lengths to purify his scheme, substituting mechanical (but imaginary) vortices for hidden (but real) forces like magnetism. Newton was rebelling against Descartes, and nowhere more fiercely than in the realm of the very small. The philosophers stood further removed from atoms than from the stars. Atoms remained a fancy, invisible to human sight. The forces governing heavenly bodies were invisible too, but ready to be inferred from a mathematical treatment of the accumulating data. For any practitioner of chemistry or alchemy, one question loomed: what made particles cohere in the first place?[19] What caused inert atoms to stick together, to form minerals and crystals and – even more wonderfully – plants and animals? The Cartesian style was recklessly ad hoc, Newton thought. It offered a different mechanical explanation for every new phenomenon: one for air, another for water, another for vinegar, yet another for sea salt – 'and so of other things: your Philosophy will be nothing else than a system of Hypotheses'.[20] Newton wanted a universal cause.

As with the question of light's true nature, he chose a narrow rhetorical path: veering past the question of whether his programme was or was not fundamentally mechanical, all reduced to particles and forces. Of light he had said, 'Others may suppose it multitudes of unimaginable small & swift Corpuscles of various sizes, springing from shining bodies at great distances, one after another, but yet without any sensible interval of time, & continually urged forward by a Principle of motion.'[21] For the rest:

God who gave Animals self motion beyond our understanding is without doubt able to implant other principles of motion in bodies which we may understand as little. Some would readily grant this may be a Spiritual one; yet a mechanical one might be showne . . .

Rather than turn away from what he could not explain, he plunged in more deeply. Dry powders refused to cohere. Flies walked on water. Heat radiated through a vacuum. Metallic particles impregnated mercury. Mere thought caused muscles to contract and dilate. There were forces in nature that he would not be able to understand mechanically, in terms of colliding billiard balls or swirling vortices. They were vital, vegetable, sexual forces – invisible forces of spirit and attraction. Later, it had been Newton, more than any other philosopher, who effectively purged science of the need to resort to such mystical qualities. For now, he needed them.

When he was not stoking his furnaces and stirring his crucibles, he was scrutinising his growing hoard of alchemical literature. By the century's end, he had created a private *Index chemicus*, a manuscript of more than a hundred pages, comprising more than five thousand individual references to writings on alchemy spanning centuries. This, along with his own alchemical writing, remained hidden long after his death.

IO

HERESY, BLASPHEMY,
IDOLATRY

The fatherless man, the fellow of the college named Trinity, turned to Christian theology with the same sleepless fervour he brought to alchemy. He started a notebook, writing Latin headings on the folios: Life of Christ; Miracles of Christ; Passion, Descent, and Resurrection. Some topics remained for ever blank; some filled and then overflowed with intense, scholarly and troubled notes. The topics that most absorbed his interest were the relation of God and Christ, the father and the son, and most of all, *De Trinitate*, Of the Trinity.[1] Here he swerved into heresy. He abjured this central dogma of his religion: three persons in one Godhead, holy and undivided. He denied the divinity of Jesus and of the Holy Ghost.

England's universities were above all else instruments of Christianity, and at each step in his Cambridge career Newton swore oaths avowing his faith. But in the seventh year of his fellowship, 1675, a further step would be required: he would take holy orders and be ordained to the Anglican clergy, or he would face expulsion. As the time approached, he realised that he could no longer affirm his orthodoxy. He would not be able to take a false oath. He prepared to resign.[2]

He believed in God, not as a matter of obligation but in the warp and weft of his understanding of nature. He believed in God eternal and infinite; a living and powerful Lord holding sway over all things; omnipresent, in bodies and filling *the space that is empty of body*.[3] He believed in God as immovable — and this belief fused with his vision, still not quite defined, of absolute space.[4] Newton's God had established the rules by which the universe operates, a handiwork that humans must strive to know. But this God did not set his clockwork in motion and abandon it.

He is omnipresent not only *virtually* but also *substantially* ... In him all things are contained and move, but he does not act on them nor they on him ... He is *always* and *everywhere* ... He is all eye, all ear, all brain, all arm, all force of sensing, of understanding, and of acting.[5]

If God was immutable, religion was not.[6] Close study fed both his faith and his heresy. He researched and wrote the history of the church again and again. He read the Scriptures literally and indulged a particular fascination with prophecy, which he saw as complex symbolism to be unravelled and interpreted. He considered this a duty. He set down a catalogue of fifteen rules of interpretation and seventy figures of prophecy. He sought the facts, dates and numbers. He calculated and then recalculated the time of the Second Coming, which he understood to be the restoration of primitive uncorrupted Christianity. He studied in detail the description of the Temple of Jerusalem, a structure of 'utmost simplicity and harmony of all its proportions',[7] and tried to reconstruct its floor plan from the long, rambling algorithms of the Hebrew Book of Ezekiel –

So he measured the length thereof, twenty cubits; and the breadth, twenty cubits, before the temple: and he said unto me, This is the most holy place. After he measured the wall of the house, six cubits; and the breadth of every side chamber, four cubits, round about the house on every side. And the side chambers were three, one over another, and thirty in order . . .

– an intricate puzzle in prose, another riddle to be deciphered. He struggled to work out the length of the ancient cubit. There seemed to be a message meant for him.

And if they be ashamed of all that they have done, shew them the form of the house, . . . and all the forms thereof, and all the ordinances thereof, and all the forms thereof, and all the laws thereof: and write it in their sight.

The very existence of the Bible in English – long opposed by the church establishment and finally authorised only a generation before Newton's birth – had inspired the Puritan cause. Vernacular versions of the Bible encouraged the laity to look into the texts and make their own interpretations. Scholars applied the new philosophical tools to Scripture. Anyone could pursue biblical inquiry as a self-directed enterprise; many tried to distinguish the pure Gospel from its medieval accretions. Ancient controversies came back to life. Newton was studying no less than the history of worship. He compared the Scriptures in the new English translation and in the ancient languages; he collected Bibles in Latin, Greek, Hebrew and French. He sought out and mastered the writings of the early fathers of the church: saints and martyrs, Athanasius and Arius, Origen, author of the *Hexapla*, Eusebius of Caesarea and Epiphanius of Constantia, and dozens more. He embroiled himself in the great controversy that tore at Christendom through the fourth century, at Nicaea and Constantinople.

The Trinity was a mystery. It defied rational explanation. It rested on a paradox that could be neither comprehended nor demonstrated: that the Son is fully human and fully divine. As a human Christ does not understand his divinity all at once. Nonetheless he is of the same being, *homoousious*, as the Father. One God: Father, Son, and Holy Spirit.

In the early fourth century, Arius, an ascetic churchman in Alexandria, led a rebellion against this doctrine. He taught that God alone is fully divine and immutable; that the Son was created, subordinate, and subject to growth and change. For this heresy Arius was excommunicated and condemned. His writings were burned. But enough survived to persuade Newton, brooding over them a millennium later, that the Trinitarians had carried out a fraud upon Christianity. The fraud had been perfected by monks and popes. The word *trinity* never appears in the New Testament. For explicit foundation in Scriptures, the orthodox looked to the First Epistle of John: 'For there are three that bear record in heaven, the Father, the Word, and the Holy Ghost: and these three are one.' Only the King James Version had the last phrase.[8] Newton's critical reading persuaded him that the original texts had been deliberately debased in support of false doctrine – a false infernal religion.[9]

In theology as in alchemy, he felt himself to be questing for ancient truths that had been perverted in the dark history of the past centuries. Knowledge had been lost, veiled in secret codes to hide it from the vulgar, distorted by blasphemers, priests and kings. He believed this to be true of mathematics, too, the language of God. In all these realms, he tried to recover words and laws once known and then lost. He had a mission. He believed he was doing God's work. 'Just as the world was created from dark Chaos through the bringing forth of the light,' he wrote in one manuscript, '. . . so our work brings forth the beginning out of black chaos and its first matter.'[10] In both alchemy and theology, he cherished secrecy just as the new philosophers in London repudiated it. No public science here: rather, meetings with anonymous confidants, barter of manuscripts, shadowy brotherhoods.

Arianism was undergoing a clandestine revival, but disbelief in the holy Trinity amounted to dangerous heresy nonetheless. By putting his arguments to paper Newton committed a crime that, if exposed, could have cost him his position and even his freedom.[11]

At the last moment, in 1675, Newton's precarious position at Cambridge was rescued. The king granted his request for a dispensation, an act that released the Lucasian professorship, in perpetuity, from the obligation to take holy orders.[12] This did not end his theological obsession. He perfected his heresy through decades of his life and millions of words. He marshalled his arguments and numbered them:

1. The [word] God is no where in the scriptures used to signify more then one of the thre persons at once.

2. The word God put absolutely without particular restriction to the Son or Holy ghost doth always signify the Father from one end of the scriptures to the other . . .

6. The son confesseth the father greater then him calls him his God, &c . . .

11. The son in all things submits his will to the will of the father. which could be unreasonable if he were equall to the father.[13]

No gulf divided Newton's theological reasoning from his physics and geometry. Logic proved that any divinity in the subordinate aspects of God remained derived from and dependent upon God. He drew a diagram:

To make this plainer suppose *a*, *b* & *c* are 3 bodies of which *a* hath gravity originally in it self by which it presseth upon *b* & *c* which are without any originall gravity but yet by the pressure of *a* communicated to them do presse downwards as much as *A* doth. Then there would be force in *a*, force in *b* & force in *c*, & yet they are not thre forces but one force which is originally in *a* & by communication/descent in *b* & *c*.[14]

He would not even label years as AD, preferring AC: Christ, but not the Lord. Jesus was more than a man but less than God. He was God's son, a mediator between God and humanity, chosen to be a prophet and messenger, and exalted to God's right hand. If we could decipher the prophecies and the messages, we would know a God of order, not chaos; of laws, not confusion. Newton plumbed both nature and history to find out God's plan. He rarely attended church.

Anger blazed through his theology; reason followed along behind. In his reading notes and 'articles' and 'points' and 'observations', his 'Short Schem of the True Religion' and his analysis of prophecies and revelations, he raged against the blasphemers. He called them fornicators – for he associated this special blasphemy with lust. 'Seducers waxing worse and worse,' he wrote, 'deceiving and being deceived – such as will not endure sound doctrine but after their own lusts heap to themselves teachers, having itching ears and turning away their ears from the truth.'[15] Monks, with their unclean thoughts, had perpetrated this corruption.

He felt Trinitarianism not just as error but as sin, and the sin was idolatry. For Newton this was the most detested of crimes. It meant serving false gods – 'that is, Ghosts or Spirits of dead men or such like beings'.[16] Kings were specially prone to it, 'kings being apt to enjoyn the honour of their dead ancestors,' declared this obsessive scholar, who, for himself, could not have been less apt to call on the honour of dead ancestors.

He had seldom returned home to Lincolnshire since the sojourn of the plague years, but in the spring of 1679 his mother succumbed to a fever. He left Cambridge and kept vigil with her over days and nights, till she died. He, the first-born son, not his half-brother or sisters, was her heir and executor, and he buried her in the Colsterworth churchyard next to the grave of his father.

II

FIRST PRINCIPLES

In the next year a comet came. In England it arose faint in the early morning sky for a few weeks in November till it approached the sun and faded in the dawn. Few saw it.

A more dramatic spectacle appeared in the nights of December. Newton saw it with naked eye on 12 December: a comet whose great tail, broader than the moon, stretched over the full length of King's College Chapel. He tracked it almost nightly through the first months of 1681.[1] A young astronomer travelling to France, Edmond Halley, a new Fellow of the Royal Society, was amazed at its brilliance.[2] Robert Hooke observed it several times in London. Across the Atlantic Ocean, where a handful of colonists were struggling to survive on a newfound continent, Increase Mather delivered a sermon, 'Heaven's Alarm to the World', to warn Puritans of God's displeasure.[3]

Halley served as a sometime assistant to a new officeholder, the Astronomer Royal. This was John Flamsteed, a clergyman and self-taught skywatcher appointed by the King in 1675, responsible for creating and equipping an observatory on a hilltop across the River Thames at Greenwich. The Astronomer's chief mission was to perfect star charts for the Navy's navigators. Flamsteed did this assiduously, recording star places with his telescope and sextant night after night, more than a thousand observations each year. Yet he had not seen the November comet. Now letters from England and Europe alerted him to it.[4]

Whatever comets were, omens or freaks, their singularity was taken for granted: each glowing visitor arrived, crossed the sky in a straight path, and departed, never to be seen again. Kepler had said this authoritatively, and what else could a culture of short collective memory believe?

But this year European astronomers recorded two: a faint predawn comet that came and went in November 1680, and a great giant that appeared a month later and dominated the skies till March. Flamsteed thought comets might behave like planets.[5] Immersed as he was in the geometry of the sky, charting the changes in celestial perspective as the earth orbited the sun, he predicted that the comet he had missed in November might yet return. He watched the sky for it. His intuition was rewarded; he spied a tail on 10 December, and the tail and head together, near Mercury, two days later. He had a friend at Cambridge, James Crompton, and he sent notes of his observations, hoping Crompton could pass them on to Newton. A fortnight later he wrote again, speculating, 'If we suppose it a consumeing substant 'tis much decayed and the Fuell spent which nourishes the blaze but I have much to say against this hypothesis however you may consider of it and Pray let me have your opinion.'[6] Newton read this and remained silent.

A month later Flamsteed tried again. 'It may seem that the exteriour coat of the Comet may be composed of a liquid . . . It was never well defined nor shewed any perfect limb but like a wisp of hay.' He was persuaded more than ever that the two comets were one. After all, he had predicted the reappearance. He struggled to explain the peculiar motion he had recorded. Suppose, he said, the sun attracts the planets and other bodies that come within its 'Vortex' – perhaps by some form of magnetism. Then the comet would approach the sun in a straight line, and this path could be bent into a curve by the pressure of the ethereal vortex.[8] How to explain its return? Flamsteed suggested a corresponding force of repulsion; he likened the sun to a magnet with two poles, one attracting and one repelling.

Finally Newton replied. He objected to the notion of magnetism in the sun for a simple reason: 'because the ☉ is a vehemently hot body & magnetick bodies when made red hot lose their virtue'. He was not persuaded that the two comets were one and the same, because his exquisitely careful measurements of their transit, and all the others he could collect – 6 degrees a day, 36 minutes a day, 3½ degrees a day – seemed to show acceleration suddenly alternating with retardation.[9] 'It is very irregular.' Even so, he diagrammed Flamsteed's proposal, the comet nearing the sun, swerving just short of it, and veering away. This he declared unlikely. Instead he suggested that the comet could have gone all the way around the sun and then returned.[10] He diagrammed this alternative, too. And he conceded a crucial point to Flamsteed's intuition: 'I can easily allow an attractive power in the ☉ whereby the Planets are kept in their courses about him from going away in tangent lines.'

He had never before said this so plainly. In the gestation of the calculus, in 1666, he had relied on tangents to curves – the straight lines from which curves veer, through the accumulation of infinitesimal changes. In laying the groundwork for laws of motion, he had relied on the tendency of all bodies to continue in straight lines. But he had also persisted in thinking of planetary orbits as a matter of balance between two forces: one pulling inward and the other, 'centrifugal', flinging outward. Now he spoke of just one force, pulling a planet away from what would otherwise be a straight trajectory.

This very conception had arrived at his desk not long before in a letter from his old antagonist Hooke. Now Secretary to the Royal Society, in charge of the *Philosophical Transactions*, Hooke wrote imploring Newton to return to the fold. He made glancing mention of their previous misunderstandings: 'Difference in opinion if such there be me thinks shoud not be the occasion of Enmity.'[1] And he asked for a particular favour: would Newton share any objections he might have to his idea, published five years before, that the motions of planets could be simply a compound of a straight-line tangent and 'an attractive motion towards the centrall body'. A straight line plus a continuous deflection equals an orbit.

Newton, just back in Cambridge after settling his mother's affairs, lost no time in composing his reply. He emphasised how remote he was from philosophical matters:

heartily sorry I am that I am at present unfurnished with matter answerable to your expectations. For I have been this last half year in Lincolnshire cumbred with concerns . . . *I have had no time to entertein Philosophical meditations* . . . And before that, I had for some years past been endeavouring to bend my self from Philosophy . . . which makes me almost wholy unacquainted with what Philosophers at London or abroad have lately been employed about . . . *I am almost as little concerned about it as one tradesman uses to be about another man's trade or a country man about learning.*[12]

Hooke's essay offered a 'System of the World'.[13] It paralleled much of Newton's undisclosed thinking about gravity and orbits in 1666, though Hooke's system lacked a mathematical foundation. All celestial bodies, Hooke supposed, have 'an attraction or gravitating power towards their own centres'. They attract their own substance and also other bodies that come 'within the sphere of their activity'. All bodies travel in a straight line until their course is deflected, perhaps into a circle or an ellipse, by 'some other effectual powers'. And the power of this attraction depends on distance.

Newton professed to know nothing of Hooke's idea. 'Perhaps you will incline the more to beleive me when I tell you that I did not before the receipt of your last letter, so much as heare (that I remember) of your Hypotheses.'[14] He threw Hooke a sop, however: an outline of an experiment to demonstrate the earth's daily spin by dropping a ball from a height. 'The vulgar' believed that, as the earth turns eastwards under the ball, the ball would land slightly to the west of its starting point, having been left behind during its fall. On the contrary, Newton proposed that the ball should land to the east. At its initial height, it would be rotating eastward with a slightly greater velocity than objects dòwn on the surface; thus it should 'outrun' the perpendicular and 'shoot forward to the east side'. For a trial, he suggested a pistol bullet on a silk line, outdoors on a very calm day, or in a high church, with its windows well stopped to block the wind.

He drew a diagram to illustrate the point. In it he allowed his imaginary ball to continue in a spiral to the centre of the earth.[15] This was an error, and Hooke pounced. Having promised days earlier to keep their correspondence private, he now read Newton's letter aloud to the Royal Society and publicly contradicted it.[16] An object falling through the earth would act like an orbiting planet, he said. It would not descend in a spiral – 'nothing att all akin to a spirall' – but rather, 'my theory of circular motion makes me suppose', continue to fall and rise in a sort of orbit, perhaps an ellipse or 'Elleptueid'.[17]

Once again Hooke had managed to drive Newton into a rage.[18] Newton replied once more and retreated to silence. Yet in their brief exchange the two men engaged as never before on the turf of this peculiar, un-physical, ill-defined thought experiment. It was 'a Speculation of noc Use yet', Hooke agreed. After all, the earth was solid, not void. They exchanged duelling diagrams.

They goaded each other into defining the terms of a profound problem. Hooke drew an ellipse.[19] Newton replied with a diagram based on the supposition that the attractive force would remain constant but also considered the case where gravity was – to an unspecified degree – greater nearer the

How a body falls to the centre of the earth:
Newton and Hooke's debate of 1679.

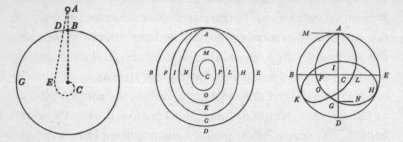

a. Newton: A body dropped from a height at A should be carried forward
by its motion and land to the east of the perpendicular, 'quite contrary to
the opinion of the vulgar'. (But he continues the path – erroneously –
in a spiral to the centre.)

b. Hooke: 'But as to the curve Line which you seem to suppose it to Desend
by... Vizt a kind of spirall... my theory of circular motion makes me suppose
it would be very differing and nothing att all akin to a spirall but rather
a kind Elleptueid.'

c. Newton: The true path, supposing a hollow earth and no resistance, would
be even more complex – 'an alternating ascent & descent'.

centre. He also let Hooke know that he was bringing potent mathematics to bear: 'The innumerable & infinitely little motions (for I here consider motion according to the method of indivisibles)...' Both men were thinking in terms of a celestial attractive force, binding planets to the sun and moons to the planets. They were writing about gravity as though they believed in it. Both now considered it as a force that pulls heavy objects down to the earth. But what could be said about the power of this force? First Hooke had said that it depended on a body's distance from the centre of the earth. He had been trying in vain to measure this, with brass wires and weights on top of St Paul's steeple and Westminster Abbey. Meanwhile the intrepid Halley, an eager seagoing traveller, had carried a pendulum up a 2,500-foot hill on St Helena, south of the equator, and judged that it swung more slowly there.

Hooke and Newton had both jettisoned the Cartesian notion of vortices. They were explaining the planet's motion with no resort to ethereal pressure (or, for that matter, resistance). They had both come to believe in a body's inherent force – its tendency to remain at rest or in motion – a concept for which they had no name. They were dancing around a pair of questions, one the mirror of the other:

What curve will be traced by a body orbiting another in an inverse-square gravitational field? (An ellipse.)

What gravitational force law can be inferred from a body orbiting another in a perfect ellipse? (An inverse-square law.)

Hooke finally did put this to Newton: 'My supposition is that the Attraction always is in a duplicate proportion to the Distance from the Center Reciprocall' – that is, inversely as the square of distance.[20] He got no reply. He tried again:

It now remaines to know the proprietys of a curve Line . . . made by a centrall attractive power . . . in a Duplicate proportion to the Distances reciprocally taken. I doubt not but that by your excellent method you will easily find out what that Curve must be, and its proprietys, and suggest a physicall Reason of this proportion.[21]

Hooke had finally formulated the problem exactly. He acknowledged Newton's superior powers. He set forth a procedure: find the mathematical curve, suggest a physical reason. But he never received a reply.

Four years later Edmond Halley made a pilgrimage to Cambridge. Halley had been discussing planetary motion in coffee houses with Hooke and the architect Christopher Wren. Some boasting ensued. Halley himself had worked out (as Newton had in 1666) a connection between an inverse-square law and Kepler's rule of periods – that the cube of a planet's distance from the sun varies as the square of its orbital year. Wren claimed that he himself had guessed at the inverse-square law years before Hooke, but could not quite work out the mathematics. Hooke asserted that he could show how to base all celestial motion on the inverse-square law and that he was keeping the details secret for now, until more people had tried and failed; only then would they appreciate his work.[22] Halley doubted that Hooke knew as much as he claimed.

Halley put the question to Newton directly in August 1684: supposing an inverse-square law of attraction toward the sun, what sort of curve would a planet make? Newton told him: an ellipse. He said he had calculated this long before. He would not give Halley the proof – he said he could not lay his hands on it – but promised to redo it and send it along.

Months passed. He began with definitions. He wrote only in Latin now, the words less sullied by everyday use. *Quantitas materiæ* – quantity of matter. What did this mean exactly? He tried: 'that which arises from its density and bulk conjointly'. Twice the density and twice the space would mean four times the amount of matter. Like weight, but *weight* would not do; he could see ahead to traps of circular reasoning. Weight would depend on gravity, and gravity could not be presupposed. So, *quantity of matter*. 'This quantity I designate under the name of body or mass.'[23] Then, *quantity of motion*: the product of velocity and mass. And *force* – innate, or impressed, or 'centripetal' – a coinage, to mean action towards a centre. Centripetal force could be absolute, accelerative, or motive. For the reasoning to come, he needed a foundation of words that did not exist in any language.

He could not, or would not, give Halley a simple answer. First he sent a treatise of nine pages, 'On the Motion of Bodies in Orbit'.[24] It firmly tied a centripetal force, inversely proportional to the square of distance, not only to the specific geometry of the ellipse but to all Kepler's observations of orbital motion. Halley rushed back to Cambridge. His one copy had become an object of desire in London. Flamsteed complained: 'I beleive I shall not get a sight of [it] till our common freind Mr Hooke & the rest of the towne have been first satisfied.'[25] Halley begged to publish the treatise, and he begged for more pages, but Newton was not finished.

As he wrote, computed, and wrote more, he saw the pins of a cosmic lock tumbling into place, one by one. He pondered comets again: if they obeyed the same laws as planets, they must be an extreme case, with vastly elongated orbits. He wrote to Flamsteed asking for more data.[26] He first asked about two particular stars, but Flamsteed guessed immediately that

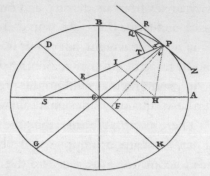

The birth of universal gravitation: Newton proves by geometry
that if a body Q orbits in an ellipse, the implied force toward the focus S
(not the centre C) varies inversely with the square of distance.

his quarry was the comet. 'Now I am upon this subject,' Newton said, 'I would gladly know the bottom of it before I publish my papers.' He needed numbers for the moons of Jupiter, too. Even stranger: he wanted tables of the tides. If celestial laws were to be established, all the phenomena must obey them.

The alchemical furnaces went cold; the theological manuscripts were shelved. A fever possessed him, like none since the plague years. He ate mainly in his room, a few bites standing up. He wrote standing at his desk. When he did venture outside, he would seem lost, walk erratically, turn and stop for no apparent reason, and disappear inside once again.[27] Thousands of sheets of manuscript lay all around, here and at Woolsthorpe, ink fading on parchment, the jots and scribbles of four decades, undated and disorganised. He had never written like this: with a great purpose, and meaning his words to be read.

Though he had dropped alchemy for now, Newton had learned from it. He embraced invisible forces. He knew he was going to have to allow planets to influence one another from a distance. He was writing the principles of philosophy. But not just that: the *mathematical* principles of *natural* philosophy. 'For the whole difficulty of philosophy,' he wrote, 'seems to be to discover the forces of nature from the phenomena of motions and then to demonstrate the other phenomena from these forces.'[28] The planets, the comets, the moon and the sea. He promised a mechanical programme – no occult qualities. He promised proof. Yet there was mystery in his forces still.

First principles. 'Time, space, place, and motion' – he wished to blot out everyday knowledge of these words. He gave them new meanings, or, as he saw it, redeemed their true and sacred meanings.[29] He had no authority to rely on – this unsocial, unpublished professor – so it was a sort of bluff, but he made good on it. He established time as independent of our sensations; he established space as independent of matter. Thenceforth *time* and *space* were special words, specially understood and owned by the virtuosi – the scientists.

Absolute, true, and mathematical time, in and of itself, and of its own nature, without reference to anything external, flows uniformly . . .

Absolute space, of its own true nature without reference to anything external, always remains homogeneous and immovable . . .[30]

Our eyes perceive only relative motion: a sailor's progress along his ship, or the ship's progress on the earth. But the earth, too, moves, in reference to *space* – itself immovable because it is purely mathematical, abstracted from our senses. Of time and space he made a frame for the universe and a credo for a new age.

12

EVERY BODY
PERSEVERES

It was ordered, that a letter of thanks be written to Mr NEWTON,' recorded Halley, as clerk of the Royal Society, on 28 April 1686, '. . . and that in the meantime the book be put into the hands of Mr HALLEY.'¹

Only Halley knew what was in 'the book' – a first sheaf of manuscript pages, copied in Cambridge by Newton's amanuensis² and dispatched to London with the grand title *Philosophiæ Naturalis Principia Mathematica*. Halley had been forewarning the Royal Society: 'a mathematical demonstration of the Copernican hypothesis'; 'makes out all the phenomena of the celestial motions by the only supposition of a gravitation towards the centre of the sun decreasing as the squares of the distances therefrom reciprocally'.³ Hooke heard him.

It was Halley, three weeks later, who undertook the letter of thanks: 'Your Incomparable treatise', etc. He had persuaded the members, none of whom could have read the manuscript, to have it printed, in a large quarto, with woodcuts for the diagrams. There was just one thing more he felt obliged to tell Newton: 'viz, that Mr Hook has some pretensions upon the invention of the rule of the decrease of Gravity . . . He sais you had the notion from him [and] seems to expect you should make some mention of him, in the preface . . ."[4]

What Newton had delivered was Book I of the *Principia*. He had completed much of Book II, and Book III lay not far behind. He interrupted himself to feed his fury, search through old manuscripts, and pour forth a thunderous rant, mostly for the benefit of Halley. He railed that Hooke was a bungler and a pretender:

This carriage towards me is very strange & undeserved, so that I cannot forbeare in stating that point of justice to tell you further . . . he should rather have excused himself by reason of his inability. For tis plain by his words he knew not how to go about it. Now is this not very fine? Mathematicians that find out, settle & do all the business must content themselves with being nothing but dry calculators & drudges & another that does nothing but pretend & grasp at all things must carry away all the invention . . .

Mr Hook has erred in the invention he pretends to & his error is the cause of all the stirr he makes . . .

He imagins he obliged me by telling me his Theory, but I thought my self disobliged by being upon his own mistake corrected magisterially & taught a Theory which every body knew & I had a truer notion of it then himself. Should a man who thinks himself knowing, & loves to shew it in correction & instructing others, come to you when you are busy, & notwithstanding your excuse, press discourses upon you & through his own mistakes correct you & multiply discourses & then make this use of it, to boast that he taught you all he spake & oblige you to acknowledge it & cry out injury & injustice if you do not, I beleive you would think him a man of a strange unsociable temper.[5]

In his drafts of Book II, Newton had mentioned the most illustrious Hooke – 'Cl[arissimus] Hookius'[6] – but now he struck all mention of Hooke and threatened to give up on Book III. 'Philosophy is such an impertinently litigious Lady that a man had as good be engaged in Law suits as have to do with her. I found it so formerly & now I no sooner come near her again but she gives me warning.'[7] Hooke had not been the first to propose the inverse-square law of attraction; anyway, for him it was a guess. It stood in isolation, like countless other guesses at the nature of the world. For Newton, it was embedded, linked, inevitable. Each part of Newton's growing system reinforced the others. In its mutual dependency lay its strength.

Halley, meanwhile, found himself entangled in the business of publishing. The Royal Society had never actually agreed to print the book. Indeed, it had only underwritten the publication of one book before, a lavish and disastrously unsuccessful two-volume *History of Fishes*.[8] After much discussion the council members did vote to order the *Principia* printed – but by Halley, at his own expense. They offered him leftover copies of *History of Fishes* in place of his salary. No matter. The young Halley was a believer, and he embraced his burden: the proof sheets mangled and lost, the complex abstruse woodcuts, the clearing of errata, and above all the nourishing of his author by cajolement and flattery. 'You will do your self the honour of perfecting scientifically what all past ages have but blindly groped after.'[9] The flattery was sincere, at least.

Halley sent sixty copies of *Philosophiæ Naturalis Principia Mathematica* on a wagon from London to Cambridge in July 1687. He implored Newton to hand out twenty to university colleagues and carry forty around to booksellers, for sale at five or six shillings apiece.[10] The book opened with a florid ode of praise to its author, composed by Halley. When an adulatory anonymous review appeared in the *Philosophical Transactions*, this, too, was by Halley.[11]

Without further ado, having defined his terms, Newton announced the laws of motion.

Law 1. Every body perseveres in its state of being at rest or of moving uniformly straight forward, except in so far as it is compelled to change its state by forces impressed. A cannonball would fly in a straight line forever, were it not for air resistance and the downward force of gravity. The first law stated, without naming, the principle of inertia, Galileo's principle, refined. Two states – being *at rest* and *moving uniformly* – are to be treated as the same. If a flying cannonball embodies a force, so does the cannonball at rest.

Law 2. A change in motion is proportional to the motive force impressed and takes place along the straight line in which that force is impressed. Force generates motion, and these are quantities, to be added and multiplied according to mathematical rules.

Law 3. To any action there is always an opposite and equal reaction; in other words, the actions of two bodies upon each other are always equal and always opposite in direction. If a finger presses a stone, the stone presses back against the finger. If a horse pulls a stone, the stone pulls the horse. Actions are interactions – no preference of vantage point to be assigned. If the earth tugs at the moon, the moon tugs back.[12]

He presented these as axioms, to serve as the foundation for an edifice of reasoning and proof. 'Law' – *lex* – was a strong and peculiar choice of word.[13] Bacon had spoken of laws, fundamental and universal. It was no coincidence that Descartes, in his own book called *Principles of Philosophy*, had attempted a set of three laws, *regulæ quædam sive leges naturæ*, specifically concerning motion, including a law of inertia. For Newton, the laws formed the bedrock on which a whole system would lie.

A law is not a cause, yet it is more than a description. A law is a rule of conduct: here, God's law, for every piece of creation. A law is to be obeyed, by inanimate particles as well as sentient creatures. Newton chose to speak not so much of God as of nature. 'Nature is exceedingly simple and conformable to herself. Whatever reasoning holds for greater motions, should hold for lesser ones as well.'[14]

Newton formed his argument in classic Greek geometrical style: axioms, lemmas, corollaries; *Q.E.D.* As the best model available for perfection in knowledge, it gave his physical programme the stamp of certainty. He proved facts about triangles and tangents, chords and parallelograms, and from there, by a long chain of argument, proved facts about the moon and the tides. On his own path to these discoveries, he had invented a new mathematics, the integral and differential calculus. The calculus and the discoveries were of a piece. But he severed the connection now. Rather than offer his readers an esoteric new mathematics as the basis for his claims, he grounded them in orthodox geometry – orthodox, yet still new, because he had to incorporate infinities and infinitesimals. Static though his diagrams looked, they depicted processes of dynamic change. His lemmas spoke of quantities that *constantly tend to equality* or *diminish indefinitely*; of areas that *simultaneously approach* and *ultimately vanish*; of *momentary increments* and *ultimate ratios* and *curvilinear limits*. He drew lines and triangles that looked finite but were meant to be on the point of vanishing. He cloaked modern analysis in antique disguise.[15] He tried to prepare his readers for paradoxes.

It may be objected that there is no such thing as an ultimate proportion of vanishing quantities, inasmuch as before vanishing the proportion is not ultimate, and after vanishing it does not exist at all . . . But the answer is easy . . . the ultimate ratio of vanishing quantities is to be understood not as the ratio of quantities before they vanish or after they have vanished, but the ratio with which they vanish.[16]

The diagrams appeared to represent space, but time kept creeping in: 'Let the time be divided into equal parts . . . If the areas are very nearly proportional to the times . . .'

When he and Hooke had debated the paths of comets and falling objects, they had dodged one crucial problem. All the earth's substance is not concentrated at its centre but spread across the volume of a great sphere – countless parts, all responsible for the earth's attractive power. If the earth as a whole exerts a gravitational force, that force must be calculated as the sum of all the forces exerted by those parts. For an object near the earth's surface, some of that mass would be down below and some would be off to the side. In later terms this would be a problem of integral calculus; in the *Principia* he solved it geometrically, proving that a perfect spherical shell would attract objects outside it exactly as by a force inversely proportional to the square of the distance to the centre.[17]

Meanwhile, he had to solve the path of a projectile attracted to this centre, not with constant force, but with a force that varies continually because it depends on the distance. He had to create a dynamics for velocities changing from moment to moment, both in magnitude and in direction, in three dimensions. No philosopher had ever conceived such a thing, much less produced it.

A handful of mathematicians and astronomers on earth could hope to follow the argument. The *Principia*'s reputation for unreadability spread faster than the book itself. A Cambridge student was said to have remarked, as the figure of its author passed by, 'There goes the man that writt a book that neither he nor anybody else understands.'[18] Newton himself said that he had considered composing a 'popular' version but chose instead to narrow his readership, to avoid disputations – or, as he put it privately, 'to avoid being baited by little smatterers in mathematicks'.[19]

Yet as the chain of proof proceeded, it shifted subtly towards the practical. The propositions took on a quality of *how to*. Given a focus, find the elliptical orbit. Given three points, draw three slanted straight lines to a fourth point. Find the velocity of waves. Find the resistance of a sphere moving through a fluid. Find orbits when neither focus is given. _Q.E.D._ gave way to _Q.E.F._ and _Q.E.I._: *that which was to be done* and *that which was to be found out.* Given a parabolic trajectory, find a body's position at an assigned time.

There was meat for observant astronomers.

On the way, Newton paused to obliterate the Cartesian cosmology, with its celestial vortices. Descartes, with his own *Principia Philosophiæ*, was his chief forebear; Descartes had given him the essential principle of inertia; it was Descartes, more than any other, whom he now wished to bury. Newton banished the vortices by taking them seriously: he did the mathematics. He created methods to compute the rotation of bodies in a fluid medium; he calculated relentlessly and imaginatively, until he demonstrated that such vortices could not persist. The motion would be lost; the rotation would cease. The observed orbits of Mars and Venus could not be reconciled with the motion of the earth. 'The hypothesis of vortices . . . serves less to clarify the celestial motions than to obscure them,' he concluded.[20] It was enough to say that the moon and planets and comets glide in free space, obeying the laws of motion, under the influence of gravity.

Book III gave *The System of the World.* It gathered together the phenomena of the cosmos. It did this flaunting an exactitude unlike anything in the history of philosophy. Phenomenon 1: the four known satellites of Jupiter. Newton had four sets of observations to combine. He produced some numbers: their orbital periods in days, hours, minutes and seconds, and their greatest distance from the planet, to the nearest thousandth of Jupiter's radius. He did the same for the five planets, Mercury, Venus, Mars, Jupiter and Saturn. And for the moon.

From the propositions established in Book I, he now proved that all these satellites are pulled away from straight lines and into orbits by a force towards a centre – of Jupiter, the sun, or the earth – and that this force varies inversely as the square of the distance. He used the word *gravitate*. 'The moon gravitates towards the earth and by the force of gravity is always drawn back from rectilinear motion and kept in its orbit.'[21] He performed an apple-moon computation with data he had lacked in Woolsthorpe twenty years before. The moon's orbit takes 27 days, 7 hours, 43 minutes. The earth measures 123,249,600 Paris feet around. If the same force that keeps the moon in orbit draws a falling body 'in our regions', then a body should fall, in one second, 15 feet, 1 inch, and 1⁷⁄₉ lines (twelfths of an inch). 'And heavy bodies do actually descend to the earth with this very force.' No one could time a falling body with such precision, but Newton had some numbers from beating pendulums, and, performing the arithmetic, he slyly exaggerated the accuracy.[22] He said he had tested gold, silver, lead, glass, sand, salt, wood, water and wheat – suspending them in a pair of identical wooden boxes from eleven-foot cords and timing these pendulums so precisely that he could detect a difference of one part in a thousand.[23]

Furthermore, he proposed, the heavenly bodies must perturb one another: Jupiter influencing Saturn's motion, the sun influencing the earth, and the sun and moon both perturbing the sea. 'All the planets are heavy towards one another.'[24] He pronounced:

It is now established that this force is gravity, and therefore we shall call it gravity from now on.

One flash of inspiration had not brought Newton here. The path to universal gravitation had led through a sequence of claims, each stranger than the last. A force draws bodies towards the centre of the earth. This force extends all the way to the moon, pulling the moon exactly as it pulls an apple. An identical force – but towards the centre of the sun – keeps the earth in orbit. Planets each have their own gravity; Jupiter is to its satellites as the sun is to the planets. And they all attract one another, in proportion to their mass. As the earth pulls the moon, the moon pulls back, adding its gravity to the sun's, sweeping the oceans around the globe in a daily flood. The force points towards the centres of bodies, not because of anything special in the centres, but as a mathematical consequence of this final claim: that every particle of matter in the universe attracts every other particle. From this generalisation all the rest followed. Gravity is universal.

Newton worked out measurements for weights on the different planets. He calculated the densities of the planets, suggesting that the earth was four times denser than either Jupiter or the sun. He proposed that the planets had been set at different distances so that they might enjoy more or less of the sun's heat; if the earth were as distant as Saturn, he said, our water would freeze.[25]

He calculated the shape of the earth – not an exact sphere, but oblate, bulging at the equator because of its rotation. He calculated that a given mass would weigh differently at different altitudes; indeed, 'our fellow countryman Halley, sailing in about the year 1677 to the island of St. Helena, found that his pendulum clock went more slowly there than in London, but he did not record the difference.'[26]

He explained the slow precession of the earth's rotation axis, the third and most mysterious of its known motions. Like a top slightly off balance, the earth changes the orientation of its axis against the background of the stars, by about one degree every seventy-two years. No one had even guessed at a reason before. Newton computed the precession as the complex result of the gravitational pull of the sun and moon on the earth's equatorial bulge.

Into this tapestry he wove a theory of comets. If gravity was truly universal, it must apply to these seemingly random visitors as well. They behaved as distant, eccentric satellites of the sun, orbiting in elongated ellipses, crossing the plane of the planets, or even ellipses extended to infinity – parabolas and hyperbolas, in which case the comet never would return.

These elements meshed and turned together like the parts of a machine, the work of a perfect mechanic, like an intricate clock, a metaphor that occurred to many as news of the *Principia* spread. Yet Newton himself never succumbed to this fantasy of pure order and perfect determinism. Continuing to calculate where calculation was impossible, he saw ahead to the chaos that could emerge in the interactions of many bodies, rather than just two or three. The centre of the planetary system, he saw, is not exactly the sun, but rather the oscillating common centre of gravity. Planetary orbits were not exact ellipses after all, and certainly not the same ellipse repeated. 'Each time a planet revolves it traces a fresh orbit, as happens also with the motion of the Moon, and each orbit is dependent upon the combined motions of all the planets, not to mention their actions upon each other,' he wrote. 'Unless I am much mistaken, it would exceed the force of human wit to consider so many causes of motion at the same time, and to define the motions by exact laws which would allow of an easy calculation.'[27]

Yet he solved another messy, bewildering phenomenon, the tides. He had assembled data, crude and scattered though they were. Samuel Sturmy had recorded observations from the mouth of the River Avon, three miles below Bristol. Samuel Colepress had measured the ebb and flow in Plymouth Harbour. Newton considered the Pacific Ocean and the Ethiopic Sea, bays in Normandy and at Pegu in the East Indies.[28] Halley himself had analysed observations by sailors in Batsha Harbour in the port of Tunking in China. None of these lent themselves to a rigorous chain of calculation, but the pattern of two high tides per twenty-five hours was clear and global. Newton marshalled the data and made his theoretical claim. The moon and sun both pull the seas; their combined gravity creates the tides by raising a symmetrical pair of bulges on opposite sides of the earth.

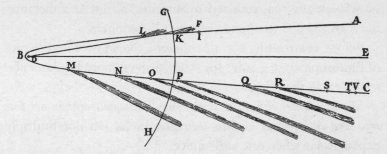

The comet of 1680 – 'as observed by Flamsteed' and 'corrected by Dr. Halley'.
Newton also collated sightings by Ponthio in Rome, Gallet in Avignon, Ango
at La Fleche, 'a young man' at Cambridge, and Mr Arthur Storer near
Hunting Creek, in Maryland, in the confines of Virginia. 'Thinking it would
not be improper, I have given . . . a true representation of the orbit which this
comet described, and of the tail which it emitted in several places.' He
concludes that the tails of comets always rise away from the sun and 'must be
derived from some reflecting matter' – smoke, or vapour.

Kepler had suggested a lunar influence on the seas. Galileo
had mocked him for it:

That concept is completely repugnant to my mind . . . I cannot bring
myself to give credence to such causes as lights, warm temperatures,
predominances of occult qualities, and similar idle imaginings . . .

I am more astonished at Kepler than at any other . . . Though he has
at his fingertips the motions attributed to the Earth, he has nevertheless lent his ear and his assent to the moon's dominion over the
waters, to occult properties, and to such puerilities.[29]

Now Newton, too, resorted to invisible action at a distance. Such arcana had to offend the new philosophers.

Before confronting the phenomena, Newton stated 'Rules of Philosophizing' – rules for science, even more fundamental in their way than the laws of motion.

No more causes of natural things should be admitted than are both true and sufficient to explain their phenomena. Do not multiply explanations when one will suffice.

The causes assigned to natural effects of the same kind must be, so far as possible, the same. Assume that humans and animals breathe for the same reason; that stones fall in America as they do in Europe; that light is reflected the same way by the earth and the planets.[30]

But the mechanical philosophy already had rules, and Newton was flouting one of them in spectacular fashion. Physical causes were supposed to be direct: matter striking or pressing on matter, not emitting invisible forces to act from afar. Action at a distance, across the void, smacked of magic. Occult explanations were supposed to be forbidden. In eliminating Descartes's vortices he had pulled away a much-needed crutch. He had nothing mechanical to offer instead. Indeed, Huygens, when he first heard about Newton's system of the world, replied, 'I don't care that he's not a Cartesian as long as he doesn't serve us up conjectures such as attractions.'[31] As a strategy for forestalling the inevitable criticism, Newton danced a two-step, confessional and defiant.

I have explained the phenomena of the heavens and of our sea by the force of gravity, but I have not yet assigned a cause to gravity

...I have not as yet been able to deduce...the reasons for these properties of gravity, and I do not feign hypotheses. For whatever is not deduced from the phenomena must be called a hypothesis; and hypotheses, whether metaphysical or physical, or based on occult qualities, or mechanical, have no place in experimental philosophy...[32]

So gravity was not mechanical, not occult, not a hypothesis. He had proved it by mathematics. 'It is enough,' he said, 'that gravity really exists and acts according to the laws that we have set forth and is sufficient to explain all the motions of the heavenly bodies and of our sea.'[33] It could not be denied, even if its essence could not be understood.

He had declared at the outset that his mission was to discover the forces of nature. He deduced forces from celestial bodies' motion, as observed and recorded. He made a great claim – the System of the World – and yet declared his programme incomplete. In fact, incompleteness was its greatest virtue. He bequeathed to science, that institution in its throes of birth, a research programme, practical and open-ended. There was work to do, predictions to be computed and then verified.

'If only we could derive the other phenomena of nature from mechanical principles by the same kind of reasoning!' he wrote. 'For many things lead me to have a suspicion that all phenomena may depend on certain forces by which the particles of bodies, by causes not yet known, either are impelled towards one another and cohere in regular figures, or are repelled from one another and recede.'[34] *Unknown* forces – as mysterious still as the forces he sought through his decades-long investigation of alchemy. His suspicion foresaw the programme of modern physics: certain forces, attraction and repulsion, final causes not yet known.

13

IS HE LIKE OTHER MEN?

As the century began Bacon had said, 'The mechanic, mathematician, physician, alchemist, and magician all immerse themselves in Nature, with a view to works, but all so far with feeble effort and slight success.'[1] He sought to prepare the stage for a new type, so far unnamed, who would interpret and penetrate nature and teach us how to command it. The prototype for *scientist* was not quite ready.

Halley heralded the *Principia* in 1687 with the announcement that its author had 'at length been prevailed upon to appear in Publick'.[2] Indeed, Newton, in his forty-fifth year, became a public man. Willy-nilly he began to develop into the eighteenth-century icon of later legend. Halley also wrote an introductory ode ('on This Splendid Ornament of Our Time and Our Nation, the Mathematico-Physical Treatise'). He sent a copy to the King – 'If ever Book was so worthy of a Prince, this, wherein so many and so great discoveries concerning the constitution of the Visible World are made out, and put past dispute, must needs be grateful to your Majesty'[3] – and for easier reading included a summary of the explanation of tides; James II had been Lord High Admiral before succeeding his brother on the throne.

'The sole Principle,' Halley explained, 'is no other than that of *Gravity*, whereby in the Earth all Bodies have a tendency toward its Center.' The sun, moon and planets all have such gravitation. The force decreases as the square of the distance increases. So a ton weight, if raised to a height of 4,000 miles, would weigh only a quarter-ton. The acceleration of falling bodies decreases in the same way. At great distances, both weight and fall become very small, but not zero. The sun's gravity is prodigious, even at the immense distance of Saturn. Thus the author with great sagacity discovers the hitherto unknown laws of the motion of comets and of the ebbing and flowing of the sea.

Truth being uniform, and always the same, it is admirable to observe how easily we are enabled to make out very abstruse and difficult matters, when once true and genuine Principles are obtained.[4]

Halley need not have bothered. James had other concerns. In his short, doomed reign, he was doing all he could to turn England towards Roman Catholicism, working his will on the army, the courts, the borough corporations and county governments, the Privy Council and – not least – the universities. In Cambridge he made an antagonist of Newton.

The King asserted his authority over this bastion of Protestantism by issuing royal mandates, placing Catholics as fellows and college officers. Tensions rose – the abhorrence of popery was written into Cambridge's statutes as well as its culture. The inevitable collision came in February 1687, when James ordered the university to install a Benedictine monk as a Master of Arts, with an exemption from the required examinations and oaths to the Anglican Church. University officials stalled and simmered. The professor of mathematics entered the fray – the resolute Puritan, theological obsessive, enemy of idolatry and licentiousness. He studied the texts: Queen Elizabeth's charter for the university, the Act of Incorporation, the statutes, the letters patent. He urged Cambridge to uphold the law and defy the King: 'Those that Councell'd his Majesty to disoblige the University cannot be his true friends ... Be courragious therefore & steady to the Laws ... If one P[apist] be a Master you may have a hundred ... An honest Courage in these matters will secure all, having Law on our sides.'[5] Before the confrontation ended, Cambridge's vice-chancellor had been convicted of disobedience and stripped of his office, but the Benedictine did not get his degree.

Newton chose a path both risky and shrewd. Cambridge's crisis was the nation's crisis in microcosm. In England's troubled soul Protestantism represented law and freedom; popery meant despotism and slavery. James's determination to Catholicise the realm led to the downfall of the House of Stuart. Within two years a Dutch fleet had invaded a divided England, James had fled to France, and a new Parliament had convened at Westminster – among its members, Isaac Newton, elected by the university senate to represent Cambridge. As the Parliament proclaimed William and Mary the new monarchs in 1689, it also proclaimed the monarchy limited and bound by the law of the land. It abolished the standing army in peacetime and established a Declaration of Rights. It extended religious toleration – except, explicitly, to Roman Catholics and to those special heretics who denied the doctrine of the Blessed Trinity. For all this Newton was present but silent. He reported back to Cambridge an argument with numbered propositions:

1. Fidelity & Allegiance sworn to the King, is only such a Fidelity & Obedience as is due to him by the law of the Land. For were that Faith and Allegiance more then what the law requires, we should swear ourselves slaves & the King absolute: whereas by the Law we are Free men . . .[6]

At the nation's hub of political power, he rented a room near the House of Commons. He put on his academic gown, combed his white hair down around his shoulders, and had his likeness painted by the most fashionable portraitist in London.[7] Word of the *Principia* was spreading in the coffee houses and abroad. He attended Royal Society meetings and social evenings. He met, and found a kind of amity with, Christiaan Huygens, now in London, and Samuel Pepys, the Royal Society's president, as well as a young Swiss mathematician and mystic, Nicolas Fatio de Duillier, and John Locke, the philosopher in most perfect harmony with the political revolution under way. Huygens still had reservations about the *Principia*'s resort to mysterious attraction, but none about its mathematical rigour, and he promoted it generously. Huygens's friend Fatio converted with loud enthusiasm to Newtonianism from Cartesianism. Fatio began serving as an information conduit between Newton and Huygens and took on the task of compiling errata for a revised edition of the

Principia. Newton felt real affection for this brash and hero-worshipping young man, who lodged with him increasingly in London and visited him in Cambridge.

Locke had just completed a great work of his own, *An Essay Concerning Human Understanding,* and saw the *Principia* as an exemplar of methodical knowledge. He did not pretend to follow the mathematics. They discussed theology – Locke amazed at the depth of Newton's biblical knowledge – and these paragons of rationality found themselves kindred spirits in the dangerous area of anti-Trinitarianism. Newton began to send Locke treatises on 'corruptions of Scripture', addressing them stealthily to a nameless 'Friend'. These letters ran many thousands of words. You seemed curious, Newton wrote, about the truth of the text of 1 John 5:7: 'the testimony of the three in heaven'. This was the keystone, the reference to *the Father, the Word, and the Holy Ghost.* Newton had traced the passage through all ages: interpretation of the Latins, words inserted by St Jerome, abuses of the Roman church, attributions by the Africans to the Vandals, variations in the margins. He said he placed his trust in Locke's prudence and calmness of temper. 'There cannot be a better service done to the truth then to purge it of things spurious,'* he said – but he nonetheless forbade Locke to publish this dangerous nonconformist scholarship.

In disputable places I love to take up with what I can best understand. Tis the temper of the hot and superstitious part of mankind in matters of religion ever to be fond of mysteries, & for that reason to like best what they understand least.

Meanwhile Pepys, who found his own mysteries in London's clubs and gaming tables, came to Newton for advice on a matter of recreational philosophy: 'the Doctrine of determining between the true proportions of the Hazards incident to this or that given Chance or Lot'. He was throwing dice for money and needed a mathematician's guidance. He asked:

A – has 6 dice in a Box, with which he is to fling a 6.
B – has in another Box 12 Dice, with which he is to fling 2 Sixes.
C – has in another Box 18 Dice, with which he is to fling 3 Sixes.
   Q. whether B & C have not as easy a Taske as A, at even luck?[9]

Newton explained why A has the best odds and gave Pepys the exact expectations, on a wager of £1,000, in pounds, shillings and pence.

All these men manoeuvred via friendly royal connections to seek a decorous and lucrative appointment for Newton in the capital. He pretended to demur – 'the confinement to the London air & a formal way of life is what I am not fond of'[10] – but these plans tempted him.

London had flourished in the quarter-century since the plague and the fire. Thousands of homes rose with walls of brick, Christopher Wren designed a new St Paul's Cathedral, streets were widened and straightened. The city rivalled Paris and Amsterdam as a centre of trading networks and a world capital of finance. England's trade and manufacturing were more centralised at one urban focus than ever before or since.

Newspapers appeared from coffee houses and printers in Fleet Street; some sold hundreds of copies. Merchants issued gazettes, and astrologers made almanacs. The flow of information seemed instantaneous compared to decades past. Daniel Defoe, recalling the plague year, wrote, 'We had no such thing as printed newspapers in those days to spread rumours and reports of things, . . . so that things did not spread instantly over the whole nation, as they do now.'[11] It was understood that knowledge meant power, even knowledge of numbers and stars. The esoteric arts of mathematics and astronomy acquired patrons greater than the Royal Society: the Navy and the Ordnance Office. Would-be virtuosi could follow periodicals that sprang into being in the eighties and nineties: *Weekly Memorials for the Ingenious* and *Miscellaneous Letters Giving an Account of the Works of the Learned*.[12]

Of the *Principia* itself, fewer than a thousand copies had been printed. These were almost impossible to find on the Continent, but anonymous reviews appeared in three young journals in the spring and summer of 1688, and the book's reputation spread.[13] When the Marquis de l'Hôpital wondered why no one knew what shape let an object pass through a fluid with the least resistance, the Scottish mathematician John Arbuthnot told him that this, too, was answered in Newton's masterwork: 'He cried out with admiration Good god what a fund of knowledge there is in that book? . . . Does he eat & drink & sleep? Is he like other men?'[14]

Its publication notwithstanding, he had never stopped working on the *Principia*. He was preparing a second edition. He scoured Greek texts for clues to his belief that the ancients had known about gravity and even the inverse-square law. He contemplated new experiments and sought new data for his complex theory of the moon's motions. Besides correcting printer's errors, he was drafting and redrafting whole new sections, refining his rules for philosophy. He struggled with the inescapable hole in his understanding of gravity's true nature. He twisted and turned: 'Tis inconceivable that inanimate brute matter should (without the mediation of something else which is not material) operate upon & affect other matter without mutual contact,' he wrote one correspondent. 'Gravity must be caused by an agent acting constantly according to certain laws, but whether this agent be material or immaterial is a question I have left to the consideration of my readers."[15]

He also pretended to leave to his readers – yet wrestled incessantly with – the Deity lurking in his margins. God informed Newton's creed of absolute space and absolute time. 'Can God be nowhere when the moment of time is everywhere?' he wrote in one of many new drafts that did not see light.[16] An active, interventionist God must organise the universe and the solar system: otherwise substance would be evenly diffused through infinite space or gathered together in one great mass. Surely God's hand could be seen in the division between dark matter, like the planets, and shining matter, like the sun. All this 'I do not think explicable by mere natural causes but am forced to ascribe it to the counsel & contrivance of a voluntary Agent."[17] He returned to his alchemical experiments, too.

*　　*　　*

Whether or not Newton was like other men, by the summer of 1693 he was eating and sleeping poorly. He had lived fifty years. He was unsettled, back and forth between the fens of Cambridgeshire and the London glare. At Cambridge his sinecure remained intact, but he scarcely taught or lectured now. In London he was angling for posts that required the king's patronage – a position at the Royal Mint, among others – but did not fully understand his own desires. He was uneasy in his relations with his new friends, tenuous though these relations were, after a life with little practice in friendship. Fatio had tormented him by falling ill and foreshadowing his own death – 'I got a grievous cold, which is fallen upon my lungs. My head is something out of order . . . If I am to depart this life I could wish my eldest brother . . . to succeed me in Your friendship' – and then by abruptly ending their relationship and returning to Switzerland.[18] (Fatio survived sixty years more.)

Sexual feelings, too, troubled Newton's nights. He had long since embraced celibacy. For this he had devised a rational programme:

The way to chastity is not to struggle directly with incontinent thoughts but to avert the thoughts by some imployment, or by reading, or meditating on other things . . .

Still, unwanted thoughts came. Ceaseless ratiocination disordered his senses.

... the body is also put out of its due temper & for want of sleep the fansy is invigorated about what ever it sets it self upon & by degrees inclines toward a delirium in so much that those Monks who fasted most arrived to a state of seeing apparitions of weomen & their shapes . . .[19]

Reclusive though he remained, rumours of Newton's mental state began to reach places where just a few years earlier his name had meant nothing: Fire had supposedly destroyed his papers. He was in a state of frenzy or melancholy or distemper. His friends had locked him away.[20] He had lost all capacity for philosophical thought.

Only Pepys and Locke knew the truth. They received accusatory, delusional, and then pitiable letters. First Newton wrote Pepys:

... for I am extremely troubled at the embroilment I am in, and have neither ate nor slept well this twelve month, nor have my former consistency of mind. I never designed to get anything by your interest, nor by King James favour, but am now sensible that I must withdraw from your acquaintance, and see neither you nor the rest of my friends any more . . .

Then Locke:

Sir –
   Being of opinion that you endeavoured to embroil me with woemen & by other means I was so much affected with it as that

when one told me you were sickly and would not live I answered twere better you were dead . . . I beg your pardon also for saying or thinking that there was a designe to sell me an office, or to embroile me. I am

your most humble & most
unfortunate Servant
Is. Newton[21]

Sex and ambition – all embroiled. Madness and genius as well; in the reputation spreading now, these imponderable qualities reinforced each other. Pepys bruited suggestive hints. 'I was loth at first dash to tell you,' he wrote one friend. He was concerned, 'lest it should arise from that which of all mankind I should least dread from him and most lament for, – I mean a discomposure in head, or mind, or both.'[22]

Yet by autumn Newton delved again into mathematical studies. He was systematising ancient geometrical analysis: especially the quadrature and construction of unruly curves. He continued to think of this work as rediscovery and restoration. After all, no one had fully plumbed the ancients' secrets. Lost manuscripts still turned up in dusty collections. There was such grandeur and purity in these old truths, which could burst into life, preserved across the millennium in Arabic as if in amber. 'The Analysis of the Ancients,' he wrote, 'is more simple more ingenious & more fit for a Geometer than the Algebra of the Moderns.'[23] Once again Newton's own studies, even when they were most innovative, were for himself alone.

With few exceptions his treatises remained in the purgatory of his private papers.

At the University of Oxford enthusiastic students (but there were few) could already hear astronomical lectures on the system of Newton.[24] Not at Cambridge, however. 'We at Cambridge, poor Wretches, were ignominiously studying the fictitious Hypotheses of the Cartesian,' one fellow recalled later.[25]

On the continent of Europe the Newtonian ideas were inspiring philosophers to frantic reformulations of their own theories. 'Vortices destroyed by Newton,' Huygens jotted. 'Vortices of spherical motion in their place.'[26] He debated mechanisms of gravity with the German mathematician and diplomat Gottfried Leibniz, who was rushing to publish his own version of planetary dynamics. 'I noticed you are in favour of a vacuum and of atoms,' Leibniz wrote. 'I do not see the necessity which compels you to return to such extraordinary entities.'[27] Newton's unmechanical gravity appalled him. 'The fundamental principle of reasoning is, *nothing is without cause*,' he wrote. 'Some conceive gravity to signify the attraction of bodies toward the bulk of the Earth, or their enticement towards it by a certain sympathy . . . He is admitting that no cause underlies the truth that a stone falls towards the Earth.'[28] It look Leibniz another year to brave an approach to Newton himself. He penned a salutation in grand style across a sheet of paper: '*illustri viro ISAACO NEUTONO*'.[29]

'How great I think the debt owed you, . . .' Leibniz began.

He mentioned that he, too, had been trying to extend geometry with a new kind of mathematical analysis, 'the application of convenient symbols which exhibit differences and sums . . . And the attempt did not go badly. But to put the last touches I am still looking for something big from you.' He confessed that he had been looking everywhere for publications by Newton. He had come across the name in a catalogue of English books, but that was a different Newton.

Besides mathematics Newton had returned to the most tortuous unfinished problem in the *Principia*: a full theory of the moon's motion. This was no mere academic exercise; given a precise recipe for predicting the moon's place in the sky, sailors with handheld astrolabes should finally be able to calculate their longitude at sea. A lunar theory should follow from Newton's theory of gravity: the ellipse of the lunar orbit crosses the earth's own orbital plane at a slant angle; the sun's attraction twists the lunar orbit, apogee and perigee revolving over a period of roughly nine years. But the force of solar gravity itself varies as the earth and moon, in their irregular dance, approach and recede from the sun. With a revised edition of the *Principia* in mind, he needed more data, and this meant calling upon the Astronomer Royal. Late in the summer of 1694 he boarded a small boat to journey down the River Thames and visit, for the first time, Flamsteed in Greenwich. He pried loose fifty lunar observations and a promise of one hundred more. Flamsteed was reluctant, and he demanded secrecy, because he considered these records his personal property. Soon Newton wanted more – syzygies and quadratures and octants, to be delivered by Flamsteed via penny post to a carrier who travelled between London and Cambridge every week. Flamsteed insisted on signed receipts. Newton cajoled Flamsteed and then pressured him. Revealing the data would make Flamsteed famous, Newton promised – 'make you readily acknowledged the most exact observer that has hitherto appeared in the world'. But the data alone would be worthless without a theory to give them meaning – 'if you publish them without such a theory . . . they will only be thrown into the heap of the observations of former astronomers'.[30] Indeed these men needed each other – Newton desperate for data that no one else in England could provide; Flamsteed desperate for any sign of gratitude or respect ('Mr Ns approbation is more to me then the cry of all the Ignorant in the world,' he wrote that winter) – and before long, they hated each other.

Two struggles continued in parallel: Newton grappled with Flamsteed and with a fiendish dynamical perturbation problem. When the astronomer complained of headaches, Newton advised him to bind his head with a garter.[31] Finally he learned that Flamsteed had let people know about the work in progress and rebuked him bitterly:

I was concerned to be publickly brought upon the stage about what perhaps will never be fitted for the publick & thereby the world put into an expectation of what perhaps they are never like to have. I do not love to be printed upon every occasion much less to be dunned & teezed by forreigners about Mathematical things or thought by our own people to be trifling away my time . . .[32]

Flamsteed spilled his agony into the margins: 'Was Mr Newton a trifler when he read Mathematicks for a sallery at Cambridge,' he railed, and then added, 'Persons thinke too well of themselves to acknowledge they are beholden to those who have furnisht them with the feathers they pride themselves in.'[33] Flamsteed took some small pleasure in reporting rumours of Newton's death: 'It served me to assure your freinds that you were in health they haveing heard that you were dead againe.' In return, for the rest of Flamsteed's life, he was a victim of Newton's implacable ruthlessness.

But Newton's fear of raising expectations was genuine. He grappled with distortions in the data caused by atmospheric refraction. The gravitational interaction of three disparate bodies did not lend itself to ready solution.

He did ultimately produce a practical formula for calculating the moon's motion: a hybrid sequence of equations and measurements that appeared first in 1702, as five Latin pages inside David Gregory's grand *Astronomiæ Elementa*. Gregory called it Newton's *theory*, but in the end Newton had omitted any mention of gravitation and buried his general picture under a mass of details. (He began: 'The Royal Observatory at Greenwich is to the West of the Meridian of Paris 2° 19'. Of Uraniburgh 12° 51' 30". And of Gedanum 18° 48'.') Halley quickly reprinted Newton's text as a booklet in English, saying, 'I thought it would be a good service to our Nation ... For as Dr. Gregory's Astronomy is a large and scarce Book, it is neither everyone's Money that can purchase it.' Halley hailed the theory's exactness and hoped to encourage people to use it, but 'the Famous Mr. Isaac Newton's Theory of the Moon' was little noted and quickly forgotten.[34]

Newton abandoned his Cambridge cloister for good in 1696. His smouldering ambition for royal preferment was fulfilled. Trinity had been his home for thirty-five years, but he departed quickly and left no friends behind.[35] As he emphatically told Flamsteed, he was now occupied by the King's business. He had taken charge of the nation's coin.

14

NO MAN IS A WITNESS IN
HIS OWN CAUSE

When the seventeenth century ended, the published work of Isaac Newton amounted to little more than three hundred copies of the *Principia*, most in England, a few scattered on the Continent. They were not much read, but scarcity made them valuable. Before a second edition was ready (in 1713, a quarter-century after first publication) a copy cost two guineas. At least one student saved his money and made a copy by hand.[1] Newton's nascent legend diffused only by word of mouth in a tiny community. When an anonymous solution to an esoteric geometry problem made its way to Holland, Johann Bernoulli announced that he recognised the solver '*ex ungue leonem*' – the lion by his claw.[2] In Berlin, Leibniz told the Queen of Prussia that in mathematics there was all previous history, from the beginning of the world, and then there was Newton; and that Newton's was the better half.[3] Tsar Peter of Russia travelled to England in 1698 eager to see several phenomena: shipbuilding, the Greenwich Observatory, the Mint, and Isaac Newton.[4]

The Royal Society was becalmed, its finances ragged, its membership dwindling. Hooke still dominated. Even living in London, Newton mostly stayed away. Yet numerical thinking was in vogue – calculation of all kinds was permeating the life of the polity – and it conjured Newton's name above all others. Mariners, architects and gamblers were understood to depend on mathematical methods. Mathematics had become a pillar raising up the glory and honour of England, 'the Academy of the Universe'.[5] John Arbuthnot published his *Essay on the Usefulness of Mathematical Learning* – a study which, he noted, seems to require 'a particular genius and turn of head, . . . few are so happy to be born with'. The incomparable Mr Newton had now discovered 'the grand secret of the whole Machine'. And he assured his readers that the world was made of *number, weight and measure* – echoing the Wisdom of Solomon as well as William Petty, the author of another new tract, *Political Arithmetick*.[6] Petty proposed the application of number to affairs of state and trade; the word *œconomick* barely existed, but he and like-minded scholars were counting what had not been counted before: populations, life expectancy, shipping tonnage, and the national income. Political arithmetic promised wonders, in a technological age:

One Man with a Mill can grind as much Corn, as twenty can pound in a Mortar; one Printer can make as many Copies, as an Hundred Men can write by hand; one Horse can carry upon Wheels, as much as Five upon their Backs; and in a Boat, or upon Ice, as Twenty.[7]

A decisive technology, and the most venerable example of standard measure, was the coin. Petty reckoned 'the whole Cash of England' at about six million pounds, circulating among perhaps six million souls, and by intricate calculation he showed that this was 'Mony sufficient to drive the Trade of the Nation'.

By the end of the century, though, England's money faced a crisis. The silver penny had been the base unit of value for a millennium; for half that time, the only unit. Now gold had joined silver in supporting a vivarium of changing species: groats, shillings, farthings, crowns, guineas. That grand new coin, the guinea, was supposed to be worth twenty shillings, but its value fluctuated unpredictably, as did the price of silver. Untold quantities of English coin were counterfeit. Even more were shrunken in weight and value: worn by decades of handling or deliberately trimmed at the edges by professional clippers, who then made bullion of their accumulated shards. So for thirty years, new machines, powered by horses and men – the mechanisms guarded as a state secret[8] – had milled a coinage with an ornamented rim to foil the clippers. A mongrel currency was the result. No one would spend a new coin willingly; these were mostly hoarded or, worse, melted down for export to France. 'Let one money pass throughout the king's dominion, and that let no man refuse,' King Edgar had said, centralising England's coinage in the tenth century. 'Let one measure and one weight be used, such as is observed in London.' No more. The melting houses and press rooms of the Mint, just inside the western rampart of the Tower of London, fell nearly silent as the 1690s began. Most coins circulating were blurry hammered silver, debased, mistrusted, and older than the hands through which they passed.

The crown called for guidance from eminent citizens, Locke, Wren and Newton among them. Wren proposed a decimal system; he was ignored. The new Chancellor of the Exchequer, Charles Montague, set a radical programme in motion: a complete recoinage – all old coins to be withdrawn from circulation. Montague had known Newton at Cambridge and with this support the king named him Warden of the Mint in April 1696, just as the recoinage began. Newton supervised an urgent industrial project, charcoal fires burning around the clock, teams of horses and men crowding in upon one another, garrisoned soldiers standing watch. It was a tumultuous time at the Tower and in London: the terms of the recoinage had strangled the supply of money essential to daily commerce and, not incidentally, effected a transfer of national wealth from the poor to the rich.

Newton grew rich himself, as Warden and then, from 1700 onward, Master. (From his first months he complained to the Treasury about his remuneration,[9] but as Master he received not only a salary of £500 but also a percentage of every pound coined, and these sums were far greater.) He found a house in Jermyn Street, bought luxurious, mainly crimson furniture,[10] engaged servants, and invited his twenty-year-old niece, Catherine Barton, the daughter of his half-sister, to live with him as housekeeper. She became renowned in London society for beauty and charm. Jonathan Swift was an admirer and frequent visitor. Within a half-decade she became the lover of Newton's patron Montague, by now the Earl of Halifax.[11]

By tradition the Mint posts offered easy income; Montague had promised Newton "'tis worth five or six hundred pounds per An, and has not too much bus'nesse to require more attendance than you may spare'.[12] Newton did not mind treating his professorship as a sinecure – he drew his Cambridge salary in absentia – but he ran the Mint until his death, with diligence and even ferocity. He was, after all, the master of melters and assayers and metallurgists who multiplied gold and silver on a scale that alchemists could only dream of. He wrestled with issues of unformed monetary theory and international currency.[13] There was nothing lofty about the requisite arithmetic, yet few could have persevered through the intricacies of accounting:

The Assaymasters weights are 1, 2, 3, 6, 11, 12 . . . The weight 12 is about 16 or 20 grains more or less as he pleases . . . His scales turn with the 128th part of a grain, that is with the 256th part of the weight 12 which answers to less then the 10th part of a penny weight . . . The Melter runs from 600 or 700 to 800 lb of late 1000 lb weight of silver in a pot & melts 3 potts a day . . . The pots shrink in the fire . . . 4 Millers, 12 horses two Horskeepers, 3 Cutters, 2 Flatters, 8 sizers one Nealer, thre Blanchers, two Markers, two Presses with fourteen labourers to pull them . . .[14]

In pursuing clippers and counterfeiters, he called on long-nurtured reserves of Puritan anger and righteousness. False coinage was a capital crime, high treason. Jane Housden and Mary Pitman, for example, were condemned (but pardoned) after having been caught with coining tools and trying to escape by dropping a parcel of counterfeit money into the Thames.[15] Newton often opposed such pardons. Counterfeiting was difficult to prove; he had himself made a Justice of the Peace and oversaw prosecutions himself, all the way to the gallows. William Chaloner not only coined his own guineas but tried to cover his tracks by accusing the Mint of making its own false money. Newton, managing a network of agents and prison informers, ensured that he was hanged. He ignored the convict's final plea:

Some body must have lost something to prove the Thiefe Some person robbd to prove the highwayman . . . Save me from being murthered O Dear Sr do this mercifull deed O my offending you has brought this upon me . . . O God my God I shall be murderd unless you wave me O I hope God will move your heart with mercy and pitty . . .[16]

Newton did not consider the uttering of bad money to be a victimless crime; he took it personally. For that matter, the crown held the Master of the Mint responsible for the weight and purity of its coinage, subject to enormous fines. At intervals he underwent the so-called Trial of the Pyx, named for the official coin chest, the pyx, protected by three independent locks and keys. A jury of the Goldsmiths' Company would test select coins 'by fire, by water by touch, or by weight or by all or by any of them', Newton noted in a memorandum he drafted and redrafted eight times.[17] Then, with solemn ceremony, it would present the King's Council with the verdict. Newton prepared carefully for these trials, carrying out his own assays. They showed that he had brought the standardisation of England's coins to new heights of exactness. For the coronation of Queen Anne, in 1702, he manufactured medals of gold and silver, for which he billed the Treasury, twice, precisely £2,485 18s. 3½d.[18] It was three years later, by Her Majesty's Special Grace, that he was knighted.

A portent of future trouble came from Leibniz, by second hand: 'to Mr. Newton, that man of great mind, my most devoted greeting' – and 'another matter, not only did I recognize that the most profound Newton's Method of Fluxions was like my differential method, but I said so . . . and I also informed others.'[19] In passing this on, the elderly mathematician John Wallis begged Newton to let some of his treasure out from the darkness. Newton was seen now as the curator of a hoard of knowledge, its extent unknown. Wallis told Newton he owed to the public his hypothesis of light and colour, which Wallis knew he had suppressed for more than thirty years, and much more – a full optical treatise. 'You say, you dare not yet publish it,' Wallis argued. 'And why not yet? Or, if not now, when then? You adde, lest I create you some trouble. What trouble now, more then at another time? . . . Mean while, you loose the Reputation of it, and we the Benefit.'

His return to the Royal Society had waited, all these years, for Hooke's exit. Hooke died in March 1703; within months Newton was chosen president. Past presidents had often been honorary, political figures. Newton seized power now and exercised it authoritatively. He quickly named his own Curator of Experiments. As president he attended almost every meeting; he commented from the chair on the reading of almost every paper.[20] He asserted control over the selection of council members. He shored up the society's sagging finances, in part from his own pocket. He imposed a rule that the royal mace be displayed when and only when he was presiding.

With Hooke dead, he also finally took Wallis's advice and released for publication his second great work – in English, rather than Latin,[21] and, more important, in prose rather than mathematics. This time he needed no editor. He had three 'books' based on his work from thirty years earlier on the nature of light and colour: the geometry of reflection and refraction; how lenses form images; and the workings of the eye and the telescope. The origin of whiteness; prisms; the rainbow. He added much more, in the form of 'Queries': queries on heat; queries on the ether; occult qualities, action at a distance, inertia. For good measure he included a pair of mathematical papers, the first he ever published. He titled the book *Opticks* – or, *a Treatise on the Reflexions, Refractions, Inflexions and Colours of Light*. He presented it to the Royal Society with an 'Advertisement' in which he explained why he had suppressed this work since 1675. The reason: 'To avoid being engaged in Disputes'.[22]

Not only had Hooke died but the world had changed. Newton's style, integrating theories with mathematical experimentation, had become familiar to philosophers, and they accepted readily the same propositions that had stirred scepticism and scorn in the 1670s. In the *Opticks* Newton described his experiments vividly and revealed far more of his working style – at least, a plausible working style – than in the *Principia*. He leaped across optical wonders as across stepping stones: from the trigonometry of refraction to the use of spectacles and mirrors; from thin transparent plates to bubbles; from the composition of the rainbow to the refraction of crystals. Much of the available data was raw and imprecise, but he shrank from nothing: friction, heat, putrefaction; the emission of light when bodies burn and when their parts vibrate. He considered the mysterious property called 'electricity' – a vapour, or fluid, or vital force that seemed to arise from the excitation of glass, or cloth, as in his 1675 experiment with bits of paper.

But was light to be understood as waves or particles? He still believed, hypothetically, that light was a stream of material particles, but he explored wavy-seeming phenomena, too: 'Do not rays of light move sometimes like an eel?' With Hooke buried, Newton also buried the ether as a medium that might vibrate with light waves, as a pond carries waves when struck by a stone. Such an ether would interfere with the planets' permanent motion, otherwise so perfectly established now.

He was committed to his corpuscular theory: that rays of light are 'very small Bodies emitted from shining Substances'.[23] Thus he seemed to take a wrong turn: over the next two centuries, researchers thrived by treating light as waves, choosing smoothness over granularity in their fundamental view of energy. The mathematical treatment of colours depended on wavelength and frequency. Until, that is, Einstein showed that light comes in quanta after all. Yet it was Newton, more than any other experimenter, who established the case for light waves. With an accuracy measured in hundredths of an inch, he had studied coloured rings in thin films.[24] He found it impossible to understand this as anything but a form of periodicity – oscillation or vibration. Diffraction, too, showed unmistakable signs of periodicity. He could neither reconcile these signs with his corpuscular theory nor omit them from his record. He could not see how a particle could be a wave, or embody waviness. He resorted to an odd word: *fits*, as in 'fits of easy reflection' and 'fits of easy transmission'. 'Probably it is put into such fits at its first emission from luminous bodies, and continues in them during all its progress. For these Fits are of a lasting nature.'[25]

Opticks stretched to cosmology and metaphysics – the more as Newton extended it in new printings. He could speak with authority now. He used his pulpit to issue a manifesto. He repeated again and again these dicta: that nature is consonant; that nature is simple; that nature is conformable to herself.[26] Complexity can be reduced to order; the laws can be found. Space is an infinite void. Matter is composed of atoms – hard and impenetrable. These particles attract one another by unknown forces: 'It is the Business of experimental Philosophy to find them out.'[27] He was charging his heirs and followers with a mission, the perfection of natural philosophy. He left them a task of further study, 'the Investigation of difficult Things by the Method of Analysis'.[28] They need only follow the signs and the method.

As President of the Royal Society he employed two new Curators of Experiments.[29] Sometimes he had them demonstrate or extend features of the *Principia* – once, for example, dropping lead weights and inflated hogs' bladders from a church tower – but more often he tried to spur experiments on light, heat and chemistry. One line of experiments explored the electric effluvium, creating a luminous glow, for example, in a glass tube rubbed with cloth, and testing the tube's attractive power with a feather. Some spirit, it seemed, could penetrate glass, move small objects, and emit light – but what? In revising the *Opticks* he drafted new 'Queries': for example, 'Do not all bodies therefore abound with a very subtle, but active, potent, electric spirit by which light is emitted, refracted, & reflected, electric attractions and fugations are performed . . . ?'[30] He suppressed these; even so, the trail of electrical research in the next century seemed to lead back to the *Opticks*.

'I have only begun the analysis of what remains to be discover'd,' he wrote, 'hinting several things about it, and leaving the Hints to be examin'd and improv'd by the farther Experiments and Observations of such as are inquisitive.'[31] Active principles – shades of alchemy – remained to be found out: the cause of gravity, of fermentation, of life itself. Only such active principles could explain the persistence and variety of motion, the constant heating of the sun and the inward parts of the earth. Only such principles stand between us and death. 'If it were not for these Principles,' he wrote,

the Earth, Planets, Comets, Sun, and all things in them, would grow cold and freeze, and become inactive Masses; and all Putrefaction, Generation, Vegetation and Life would cease.[32]

Word of the *Opticks* spread slowly through Europe; then a bit faster after a Latin edition appeared in 1706.[33] Father

Nicolas Malebranche, aging theologian and Cartesian, reviewed the *Opticks* with the remark, 'Though Mr. Newton is no physicist, his book is very interesting . . .'[34] Rivals who had never managed to dispute his mathematics found new opportunities in his metaphysics. He had spoken of infinite space as the 'sensorium' of God, by which he meant to unify omnipresence and omniscience. God, being everywhere, is immediately and perfectly aware. But the difficult word, suggesting a bodily organ for divine sensation, left him vulnerable to theological counterattack: 'I examined it and laughed at the idea,' Leibniz told Bernoulli – these eminent admirers now turned enemies of Newton. 'As if God, from whom everything comes, should have need of a sensorium. This man has little success with Metaphysics.'[35] And again Leibniz abhorred Newton's vacuum. A world of vast emptiness – unacceptable. Planets attracting one another across this emptiness – absurd. He objected to Newton's conception of absolute space as a reference frame for analysing motion, and he mocked the idea of gravitation. For one body to curve round another, with nothing pushing or impelling it – impossible. Even *supernatural.* 'I say, it could not be done without a miracle.'[36]

By now he and Newton were in open conflict. Leibniz, four years Newton's junior, had seen far more of the world – a stoop-shouldered, tireless man of affairs, lawyer and diplomat, cosmopolitan traveller, courtier to the House of Hanover. The two men had exchanged their first letters – probing and guarded – in the late 1670s. In the realm of mathematics, it was paradoxically difficult to stake effective claims to knowledge without disclosure. One long letter from Newton, for Leibniz via Oldenburg, asserted possession of a 'twofold' method for solving inverse problems of tangents 'and others more difficult' and then concealed the methods in code:

At present I have thought fit to register them both by transposed letters ... *5accdæ10effh11i4l3m9n6oqqr8s11t9v3x: 11ab3cdd10eægoillrm7n603p3 q6r5s11t8vx, 3acæ4eghs5i414m5n8oq4r3s6t4vaaddæeeeeeiijmmnnooprrssss sttuu.*[37]

He retained the key in a dated 'memorandum' to himself. Still, impenetrable though this cryptogram was, Newton had shown Leibniz powerful methods: the binomial theorem, the use of infinite series, the drawing of tangents, and the finding of maxima and minima.

Leibniz, in his turn, chose not to acknowledge these when, in 1684 and 1686, he published his related mathematical work as 'A New Method for Maxima and Minima, and Also for Tangents, Which Stops at Neither Fractions nor Irrational Quantities, and a Singular Type of Calculus for These' in the new German journal *Acta Eruditorum*. He offered rules for computing derivatives and integrals, and an innovative notation: dx, $f(x)$, $\int x$. This was a pragmatic mathematics, a mathematics without proof, an algorithm for solving 'the most difficult and most beautiful problems'.[38] With this new name, *calculus*, it travelled slowly towards England, just before word of the *Principia*, with its classic geometrical style concealing new tools of analysis, made its way across the Continent.

Now, decades later, Newton had a purpose in publishing his pair of mathematical papers with the *Opticks*, and he made

Communicating with Leibniz: the key to the cryptogram.

his purpose plain. In particular, 'On the Quadrature of Curves' laid out for the first time his method of fluxions. In effect, despite the utterly different notation, this was Leibniz's differential calculus. Where Leibniz worked with successive differences, Newton spoke of rates of flow changing through successive moments of time. Leibniz was chunklets – discrete bits. Newton was the continuum. A deep understanding of the calculus ultimately came to demand a mental bridge from one to the other, a translation and reconciliation of two seemingly incompatible symbolic systems.

Newton declared not only that he had made his discoveries by 1666 but also that he had described them to Leibniz. He released the correspondence, anagrams and all.[39] Soon an anonymous counterattack appeared in *Acta Eruditorum* suggesting that Newton had employed Leibniz's methods, though calling them 'fluxions' instead of 'Leibnizian differences'. This anonymous reviewer was Leibniz. Newton's disciples fired back in the *Philosophical Transactions*, suggesting that it was Leibniz who, having read Newton's description of his methods, then published 'the same Arithmetic under a different name and using a different notation'.[40] Between each of these thrusts and parries, years passed. But a duel was under way. Partisans joined both sides, encouraged by tribal loyalties more than any real knowledge of the documentary history. Scant public record existed on either side.

The principals joined the fray openly in 1711. A furious letter from Leibniz arrived at the Royal Society, where it was read aloud and 'deliver'd to the President to consider the contents thereof'.[41] The society named a committee to investigate 'old letters and papers'.[42] Newton provided these. Early correspondence with John Collins came to light; Leibniz had seen some of it, all those years before. The committee produced a document without precedent: a detailed, analytical history of mathematical discovery. No clearer account of the calculus existed, but exposition was not the point; the report was meant as a polemic, to condemn Leibniz, accusing him of a whole congeries of plagiarisms. It judged Newton's method to be not only the first – 'by many years' – but also more elegant, more natural, more geometrical, more useful, and more certain.[43] It vindicated Newton with eloquence and passion, and no wonder: Newton was its secret author.

The Royal Society published it rapidly. It also published a long assessment of the report, in the *Philosophical Transactions* – a diatribe, in fact. This, too, was secretly composed by Newton. Thus he anonymously reviewed his own anonymous report, and in doing so he spoke of candour:

It lies upon [Leibniz], in point of Candor, to tell us what he means by pretending to have found the Method before he had found it.

It lies upon him, in point of Candor, to make us understand that he pretended to this Antiquity of his Invention with some other Design than to rival and supplant Mr. Newton.

When he wrote those Tracts he was but a Learner, and this he ought in candour to acknowledge.

He declared righteously: 'no Man is a Witness in his own Cause. A Judge would be very unjust, and act contrary to the Laws of all Nations, who should admit any Man to be a Witness in his own Cause.'[44]

Newton wrote many private drafts about Leibniz, often the same ruthless polemic again and again, varying only by a few words. The priority dispute spilled over into the philosophical disputes, the Europeans sharpening their accusation that his theories resorted to miracles and occult qualities. What reasoning, what causes, should be permitted? In defending his claim to first invention of the calculus, Newton stated his rules for belief, proposing a framework by which his science – any science – ought to be judged. Leibniz observed different rules. In arguing against the miraculous, the German argued theologically. By pure reason, for example, he argued from the perfection of God and the excellence of his workmanship to the impossibility of the vacuum and of atoms. He accused Newton – and this stung – of implying an imperfect God.

Newton had tied knowledge to experiments. Where experiments could not reach, he had left mysteries explicitly unsolved. This was only proper, yet the Germans threw it back in his face: 'as if it were a Crime to content himself with Certainties and let Uncertainties alone'.

'These two Gentlemen differ very much in Philosophy,' Newton declared under cover of anonymity.

The one teaches that Philosophers are to argue from Phænomena and *Experiments* to the Causes thereof, and thence to the Causes of those Causes, and so on till we come to the first Cause; the other that all the Actions of the first Cause are Miracles, and all the Laws imprest on Nature by the Will of God are perpetual Miracles and occult Qualities, and therefore not to be considered in Philosophy. But must the constant and universal Laws of Nature, if derived from the Power of God or the Action of a Cause not yet known to us, be called Miracles and occult Qualities?[245]

Newton understood the truth full well: that he and Leibniz had created the calculus independently. Leibniz had not been altogether candid about what he had learned from Newton – in fragments, and through proxies – but the essence of the invention was his. Newton had made his discoveries first, and he had discovered more, but Leibniz had done what Newton had not: published his work for the world to use and to judge. It was secrecy that spawned competition and envy. The plagiarism controversy drew its heat from the gaps in the dissemination of knowledge. In a young and suddenly fertile field like the mathematics of the seventeenth century, discoveries had lain waiting to be found again and again by different people in different places.[46]

The Newton-Leibniz duel continued long after the deaths of the protagonists. It constricted the development of English mathematics, as orthodoxy hardened around Newton's dot notation.[47] The more historians came to understand what happened, the uglier it looked. No one could dispute Lenore Feigenbaum's simple précis: 'Grown men, brilliant and power-ful, betrayed their friends, lied shamelessly to their enemies, uttered hateful chauvinistic slurs, and impugned each others' characters.'[48] Newton's rage, Leibniz's bitterness – the darkest emotions of these protoscientists almost overshadowed their shared achievement.

Yet the priority dispute contributed to the transition of sci-ence from private obsessions to public enterprise. It exposed texts that Newton had meant to keep hidden and concentrated the interest of philosophers in these new methods: their rich-ness, their fungibility, their power. The competition between formalisms – superficially so different – brought into focus the shared underlying core.

The obsessions of Newton's later years disappointed modernity in some way. Later Newtonians came to find them as troubling as his pursuit of alchemy and biblical prophecy, if not for quite the same reasons. Just when science began to coalesce as an English institution, Newton made himself its autocrat. He purged the Royal Society of all remnants of Hooke. He gained authority over the Observatory and wrested from Flamsteed the astronomer's own life's work, a com-prehensive catalogue of the stars. (Flamsteed, summoned to appear before Newton, 'complained then of my catalogue being printed by Halley, without my knowledge, and that I was robbed of the fruits of my labors. At this he fired, and called me all the ill names, puppy &c. that he could think of.'[49]) D. T. Whiteside, who became the twentieth century's pre-eminent scholar and shepherd of Newton's mathematical work, could not but remark:

Only too few have ever possessed the intellectual genius and surpassing capacity to stamp their image upon the thought of their age and that of centuries to follow. Watching over the minting of a nation's coin, catching a few counterfeiters, increasing an already respectably sized personal fortune, being a political figure, even dictating to one's fellow scientists: it should all seem a crass and empty ambition once you have written a *Principia*.

Still, it did not seem so to Newton.[50] He had been a man on God's mission, seeking his secrets, interpreting his design, but he had never meant to draw philosophers to his side. He had not meant to lead a cult or a school. Nevertheless he had gathered disciples and enemies as well. Leibniz never stopped hoping for a moral victory. *Adieu*, he wrote. 'Adieu the vacuum, the atoms, and the whole Philosophy of M. Newton.'[51]

Leibniz died in 1716, having spent his last years at Hanover as librarian to the Duke. Newton's death was still to come.

15

THE MARBLE INDEX
OF A MIND

News came swiftly from far and exotic lands. *Philosophical Transactions* reported the discovery of 'Phillippine-Islands' and 'Hottentots'.[1] Thus inspired, in 1726 a Fleet Street printer produced a volume of *Travels into Several Remote Nations of the World*, by one Captain Lemuel Gulliver, describing wonderful peoples: Yahoos and Brobdingnagians. At length Gulliver's travels brought him to Glubbdubdrib, the island of sorcerers, where he heard the ancients and the moderns compare their histories.[2] Aristotle appeared, with lank hair and meagre visage, confessed his mistakes, noted that Descartes's vortices were also soon 'to be exploded', and offered up some epistemological relativism:

He predicted the same fate to ATTRACTION, whereof the present learned are such zealous asserters. He said, 'that new systems of nature were but new fashions, which would vary in every age; and even those, who pretend to demonstrate them from mathematical principles, would flourish but a short period of time, and be out of vogue when that was determined.'

The shade of Aristotle might think so. Never had human cosmologies come and gone so rapidly, the new sweeping aside the old in scarcely a lifetime. Jonathan Swift had no reason to know that Newton's would be the one to endure.

It scarcely mattered, Voltaire said cynically. Hardly anyone knew how to read, and of these few, hardly any read philosophy. 'The number of those who think is exceedingly small, and they are not interested in upsetting the world.'³ Nevertheless, captivated by Newtonianism, he began to spread the word in his own writing – popular science and myth-making. He told the story of the apple, which he had heard from Newton's niece. 'The labyrinth and abyss of infinity is another new journey undertaken by Newton and he has given us the thread with which we can find our way through.' And he defended Newton from the many French accusers, 'learned or not', who complained of his replacing familiar *impulsion* with mysterious *attraction*. He conjured a reply in Newton's voice:

You no more understand the word impulsion than you do the word attraction, and if you cannot grasp why one body tends towards the centre of another, you cannot imagine any the more by what virtue one body can push another . . . I have discovered a new property of matter, one of the secrets of the Creator. I have calculated and demonstrated its effects; should people quibble with me over the name I give it?⁴

Other memorialists of Newton in England and Europe put on record personal details, of a certain kind. The great man had clear eyesight and all his teeth but one. He had kept a head of pure white hair. He remained gentle and modest, treasuring quiet and disliking squabbles. He never laughed – except once, when asked what use in life was reading Euclid, 'upon which Sir Isaac was very merry'. He had died, from a stone in his bladder, after hours of agony, sweat rolling from his forehead, but he had never cried out or complained.[5]

In England, where new popular gazettes carried curiosities to the countryside, the death of Newton inspired a decade-long outpouring of verse, patriotic and lyrical. He was after all the philosopher of light. Elegists seemed to give him credit for all the colours he had found in his prism, flaming red, tawny orange, deepened indigo. Richard Lovatt posted a poem to the *Ladies Diary* in 1733:

> . . . mighty Newton the Foundation laid,
> Of his Mysterious Art . . .
> Great Britain's sons will long his works pursue.
> By curious Theorems he the Moon cou'd trace
> And her true Motion give in every Place.[6]

A hero, an English hero, and a new kind of hero, brandishing no sword but 'curious theorems'. The connection between knowledge and power had been made. Not all forms of knowledge were equal: the *Gentleman's Magazine* complained about schools 'where the two chief branches of Knowledge inculcated are French and Dancing', but reported with pleasure that a medal honouring Newton had been struck at the

Tower.[7] More poetry followed; an enthusiast could bring off a
paean in just two lines:

> Newton's no more – By Silence Grief's exprest:
> Lo here he lies; His World proclaim the rest.[8]

Alexander Pope's couplet found more readers:

> Nature and Nature's laws lay hid in night;
> God said, *Let Newton be!* And All was *Light.*[9]

Public lectures and travelling demonstrations went where
the written word lacked force. Newton had made claims that
could be tested. By computation he pronounced the earth
oblate, broader at the equator, in contrast to the egg-shaped
Cartesian earth. In 1733 the French Academy of Sciences proposed to settle the matter and dispatched expeditions north-
ward to Lapland and southward to Peru with quadrants,
telescopes and twenty-foot wooden rods. When the voyagers
returned – a decade later – they brought measurements
supporting Newton's view. Mastery of the stars and planets
empowered the nation's ships as much as the wind did. Halley
showed by example what it meant to believe in Newtonianism.
He made dramatic public predictions, computing the path of a
certain comet and prophesying its return every seventy-six
years; the forecast in itself inspired and disturbed the English
long before it proved true. In 1715 Halley anticipated a total
solar eclipse by publishing a broadsheet map showing where
and when the moon's shadow would cross England. The Royal

Society gathered at the appointed moment in a courtyard and on a rooftop, under a clear sky, where they saw the sudden untimely nightfall, the sun's corona flaring, and owls, confused, taking to the air. They saw that by predicting celestial prodigies an astronomer tamed them and drained them of their terror.[10]

As it evolved into a new orthodoxy, Newtonianism became a target. It was continually being disproved, in tracts with titles like *Remarks upon the Newtonian philosophy: wherein the fallacies of the pretended mathematical demonstrations, by which those authors support that philosophy are clearly laid open: and the philosophy itself fully proved to be false and absurd both by mathematical and physical demonstration.*[11] It inspired satires, some deliberate and some ingenuously respectful. One Newtonian convert, the vicar of Gillingham Major, wrote a treatise called *Theologiæ Christianæ Principia Mathematica*, calculating that the probability of counterevidence to the Gospels diminished with time and would reach zero in the year 3144. A Viennese physician, Franz Mesmer, 'discovered' animal magnetism or animal gravity, a healing principle based (so he claimed) on Newtonian principles. He named it after himself: Mesmerism.

But Newtonianism was not yet a word, in English.[12] In Italy, an instructive little tract appeared with the title *Il Newtonianismo per le Dame*, quickly rendered into French and then English as *Sir Isaac Newton's Philosophy Explain'd for the Use of the Ladies*, in six dialogues, vivid and heroic. It employed the inverse-square law to calculate the power of attraction between separated lovers. And the philosopher wielded a sword after all: 'Thus Sir Isaac Newton, the avowed Enemy to imaginary Systems, and to whom you are indebted for the true idea of Philosophy, has at one Blow lopped off the two principal Heads of the reviving Cartesian Hydra.'[13]

That heroic style went out of vogue soon enough. Now poets do not glorify Newton, but they can love him, or his legend. 'Maybe he made up the apple,/Maybe not,' ventures Elizabeth Socolow:

> I see the way he thirsted all his life
> to find the force that seemed not to be there,
> but acted, and precisely.[14]

For centuries between, the poets doubted him and even demonised him – his calculating spirit, his icy rationality, his plundering of the mysteries *they* owned. Then Newton was created as much by his enemies as his friends.

Keats and Wordsworth joined the Romantic artist Benjamin Haydon at dinner on a bleak December night in 1817 in his painting-room.[15] He showed them his broad, unfinished canvas of *Christ's Entry into Jerusalem*; in the crowd of Christ's followers he had painted the face of Newton. Keats ragged him for that and proposed a sardonic toast: 'Newton's health, and confusion to mathematics.' Newton had unweaved the rainbow with his prism. He had reduced nature to philosophy; had made knowledge a 'dull catalogue of common things'; had tried to 'conquer all mysteries by rule and line'.[16] Shelley complained that, to Newton,

> Those mighty spheres that gem infinity
> Were only specks of tinsel fixed in heaven
> To light the midnights of his native town![17]

He could not acknowledge that it was Newton for whom the stars had grown to mighty spheres. Wordsworth, too, had an image in mind, cold yet majestic. He saw at Trinity College a statue in the moon's light:

> Newton with his prism and silent face,
> The marble index of a mind for ever
> Voyaging through strange seas of Thought, alone.[18]

Loathing Newton most profoundly was the myth-maker William Blake, poet, engraver, and visionary. Blake was born to hate Newton. He loathed him and revered him. When he drew Newton he pictured a demigod, naked and muscular, with golden locks and keen hands. But he also saw an enemy of imagination: the lawmaker and repressor – 'unknown, abstracted, brooding, secret, the dark Power hid'.[19] Like Leibniz and the Cartesians he feared Newton's vacuum; unlike them, he believed in it: 'this abominable Void, this soul-shudd'ring Vacuum'. He blamed Newton for perfection and rigidity. He blamed him for his very success as a truthseeker. 'God forbid that Truth should be Confined to Mathematical Demonstration.'[20] He blamed him for departing from the particular by abstraction and generalisation. He blamed him for the reason that trumps imagination, and he blamed him for finding knowledge by way of doubt:

> Reason says Miracle; Newton says Doubt
> Aye thats the way to make all nature out
> Doubt Doubt & dont believe without experiment.[21]

He blamed him for the part he had played – the Romantics began to see this – in the greying of Eden, the industrialisation and mechanisation; factories dimming the air with smoke. Dark Satanic mills. 'The Water-wheels of Newton,' Blake cried:

> Of many Wheels I view, wheel without wheel, with cogs tyrannic
> Moving by compulsion each other, not as those in Eden, which
> Wheel within Wheel, in freedom revolve in harmony & peace.[22]

Newton had given, and he had taken away. He gave a sense of order, security and lawfulness. The American Declaration of Independence found Newtonianism, via Locke, and threw it back at the British by citing the laws of nature in its opening sentence. He gave infinite space yet took away the plenitude, for with infinity came the void. He took away mystery, and for some that meant godliness. An ad hoc universe had also been a providential universe.

He was made in myth, this Newton of the poets. No one tried reading the vast storehouse of paper that survived him. The manuscripts, fragmentary drafts, scraps of calculation and speculation, all lay through the generations in the private storerooms of English aristocratic families. The anti-Trinitarian heresies were rumoured but still secret. A full century passed before anyone attempted a real biography: the pious David Brewster, who in 1831 honoured the nobility of Newton's genius, emphasised his simplicity, humility and benevolence, and, though he had seen some of the disturbing manuscripts, declared firmly, 'There is no reason to suppose that Sir Isaac Newton was a believer in the doctrines of alchemy.'[23]

Brewster also stayed clear of the apple, though he had heard the story and paid a visit to the surviving tree at Woolsthorpe. It remained for the poets to ensure the apple's place in the Newton legend. They knew the apple's ancient pull: sin and knowledge; knowledge and inspiration. 'Man fell with apples, and with apples rose,' Byron wrote –

for we must deem the mode
In which Sir Isaac Newton could disclose
Through the then unpaved stars the turnpike road,
A thing to counterbalance human woes;
For ever since immortal man hath glowed
With all kinds of mechanics, and full soon
Steam-engines will conduct him to the Moon.[24]

Success bred confidence. Law triumphed. Newton's followers and successors created a more perfect Newtonianism than his own, striving for extremes of rational determinism. In post-Revolutionary France, Pierre Simon de Laplace re-expressed Newton's mechanics in a form suitable for modern field theories – rates of change as gradients and potentials – and then reached for another kind of philosopher's stone. He imagined a supreme intelligence, a perfect computer, armed with data representing the positions and forces of all things at one instant. It need only apply Newton's laws: 'Such an intelligence would embrace in the same formula the motions of the greatest bodies of the universe and those of the lightest atom; nothing would be uncertain, and the future, like the past, would be present to its eyes.'

Philosophers no longer claim him as one of their own. Philosophy absorbed him, beginning with Immanuel Kant, who turned the German tide against Leibniz and his chains of reasoning, theistic proofs, circles of words. Kant saw science as specially successful, knowledge that begins with experience. He brought space and time into epistemology; space as magnitude, empty or not; time as another kind of infinitude; both existing outside ourselves, eternal and subsistent. To explore how we know anything, we begin with our knowledge of these absolutes. Yet afterwards, Newton became a quaint figure for philosophers. When Edwin Arthur Burtt wrote his 1924 *Metaphysical Foundations of Modern Physical Science*, he first assigned those foundations to Newton and then said, without irony: 'In scientific discovery and formulation Newton was a marvellous genius; as a philosopher he was uncritical, sketchy, inconsistent, even second-rate.' He added in passing, 'It has, no doubt, been worth the metaphysical barbarism of a few centuries to possess modern science.'[25]

The *Principia* marked a fork in the road: thenceforth science and philosophy went separate ways. Newton had removed from the realm of metaphysics many questions about the nature of things – about what exists – and assigned them to a new realm, physics. 'This preparation being made,' he declared, 'we argue more safely.'[26] And less safely, too: by mathematising science, he made it possible for its facts and claims to be proved wrong.[27] This vulnerability was its strength. By the early nineteenth century Georges Cuvier was asking enviously, 'Should not natural history also one day have its Newton?' By the early twentieth, social scientists, economists, and biologists, too, were longing for a Newton of their own – or for the unattainable mirage of Newtonian perfection.[28]

Then science seemed to reject that same perfection: the absolutes and the determinism. The relativity of Einstein appeared as a revolutionary assault on absolute space and time. Motion distorts the flow of time and the geometry of space, he found. Gravity is not just a force, ineffable, but also a curvature of space-time itself. Mass, too, had to be redefined; it became interchangeable with energy.[29] George Bernard Shaw declared to radio listeners that Newtonianism had been a religion, and now it had 'crumpled up and was succeeded by the Einstein universe'.[30] T. S. Kuhn, in asserting his famous theory of scientific revolutions, said that Einstein had returned science to problems and beliefs 'more like those of Newton's predecessors than of his successors'.[31] These, too, were myths.

We understand space and time, force and mass, in the Newtonian mode, long before we study them or read about them. Einstein did shake space-time loose from pins to which Newton had bound it, but he lived in Newton's space-time nonetheless: absolute in its geometrical rigour and its independence of the world we see and feel. He happily brandished the tools Newton had forged. Einstein's is no everyday or psychological relativity.[32] 'Let no one suppose,' he said in 1919, 'that the mighty work of Newton can really be superseded by this or any other theory. His great and lucid ideas will retain their unique significance for all time as the foundation of our whole modern conceptual structure in the sphere of natural philosophy.'[33] The observer whom Einstein and his followers returned to science scarcely resembled the observer whom Newton had removed. That medieval observer had been careless and vague; time was an accumulation of yesterdays and tomorrows, slow and fast, nothing to be measured or relied upon. Time and space had first to be rescued – made absolute, true, and mathematical: *The common people conceive those quantities under no other notions but from the relation they bear to sensible objects.* Sensible meant crude – wooden measuring sticks and clocks that told only the hour. *And thence arise certain prejudices for the removing of which it will be convenient to distinguish them into absolute and relative, true and apparent, mathematical and common.* The day, as measured by successive southings of the sun, varied in length; philosophy needed an unqualified measure. It was not only convenient but necessary, in creating physics, to abstract this pure sense of time and space. Even so, Newton left openings for the relativists who followed three centuries behind. *It may be, that there is no such thing as an equable motion, whereby time may be accurately measured,* he wrote. *It may be that there is no body really at rest, to which the places and motions of others may be referred.*[34]

His insistence on a particle view of light did not lead to the modern quantum theory, even if, in some sense, it proved correct. It was Einstein who discovered the equivalence of mass and energy; still, Newton suspected their organic unity: 'Are not gross Bodies and Light convertible into one another, and may not Bodies receive much of their Activity from the Particles of Light which enter their Composition?'[35] He never spoke of *fields* of force, but field theories were born in his view of gravitational and magnetic forces distributed about a centre: 'an endeavour of the whole directed towards a centre, . . . a certain efficacy diffused from the centre through each of the surrounding places.'[36] Newton also anticipated the existence of subatomic forces by rejecting alternative explanations for the cohesion of matter: 'some have invented hooked Atoms, which is begging the Question'. Let others resort to occult qualities. 'I had rather infer from their Cohesion, that their Particles attract one another by some Force, which in immediate Contact is exceeding strong.'[37] He speculated that such a force – another force, independent of gravity, magnetism and electricity – might prevail only at the smallest distances.

The infinities, the void, the laws must endure – not a fashion, not reversible. We internalise the essence of what he learned. A few general principles give rise to all the myriad properties and actions of things. The universe's building blocks and laws are everywhere the same.[38]

No one feels the burden of Newton's legacy, looming forwards from the past, more than the modern scientist. A worry nags at his descendants: that Newton may have been too successful; that the power of his methods gave them too much authority. His solution to celestial dynamics was so thorough and so precise – scientists cannot help but seek the same exactness everywhere. 'A slightly naughty thought can come to one's mind here,' said Hermann Bondi. 'The tools that he gave us stand at the root of so much that goes on now . . . We may not be doing a lot more than following in his footsteps. We may still be so much under the impression of the particular turn he took . . . we cannot get it out of our system.'[39] We cannot. What Newton learned entered the marrow of what we know without knowing how we know it.

His papers began to appear in the early twentieth century, when cash-poor nobility sold them at auction and they scattered to collectors in Europe and across the Atlantic. In 1936 Viscount Lymington, a descendant of Catherine Barton, sent

Sotheby's a metal trunk containing manuscripts of three million words, to be broken up and offered at auction in 329 lots. Interest was slight,[40] but the economist and Cantabrigian John Maynard Keynes, disturbed, as he said, by the impiety, managed to buy some at the auction and then gradually reassembled more than a third of the collection. What he found there amazed him: the alchemist; the heretical theologian; not the cold rationalist Blake had so despised but a genius more peculiar and extraordinary. An 'intense and flaming spirit'. With the papers Keynes also bought Newton's death mask – eyeless, scowling. At least twenty portraits of Newton had been painted, not all from life; they differ extravagantly, one from another.

'Newton was not the first of the age of reason,' Keynes told a few students and fellows in a shadowed room at Trinity College. 'He was the last of the magicians, the last of the Babylonians and Sumerians, the last great mind which looked out on the visible and intellectual world with the same eyes as those who began to build our intellectual inheritance rather less than 10,000 years ago.'[41] The Newton of tradition, the 'Sage and Monarch of the Age of Reason', had to arise later.

He had concealed so much, till the very end. As his health declined, he kept writing. His niece's new husband, John Conduitt, saw him in his last days working in near darkness on an obsessional history of the world – he wrote at least a dozen drafts – *The Chronology of Ancient Kingdoms Amended*.[42] He measured the reigns of kings and the generations of Noah, used astronomical calculations to date the sailing of the Argonauts, and declared the ancient kingdoms to be hundreds of years

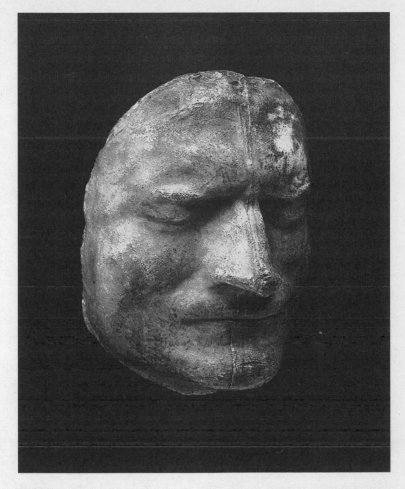

younger than generally supposed. He incorporated his analysis of the Temple of Solomon and said enough about idolatry and the deification of kings to raise suspicion of his heretical beliefs, but he suppressed those one last time.

In his chambers, after a painful fit of gout, he sat with Conduitt before a wood fire and talked about comets. The sun needed constant replenishment, he said. Comets must provide it, feeding the sun like logs thrown on the fire. The comet of 1680 had come close, and it would return. He said that on one approach, perhaps after five or six more orbits, it would fall into the sun and fuel a blaze to consume the very earth, and all its inhabitants would perish in the flames.[43] Yet, Newton said, this was mere conjecture.

He wrote: 'To explain all nature is too difficult a task for any one man or even for any one age. Tis much better to do a little with certainty & leave the rest for others that come after you.'[44] This sheet of paper, too, he abandoned.

On his deathbed he refused the sacrament of the church. Nor could a pair of doctors ease his pain. He died early Sunday morning, 19 March 1727. On Thursday the Royal Society recorded in its Journal Book, 'The Chair being Vacant by the death of Sir Isaac Newton there was no Meeting this Day.'

His recent forebears had used scriveners to draft wills directing the disposition of their meagre possessions, principally sheep. When they did not leave such documents, even their names vanished. An early chronicler, researching Newton's story soon after his death, delved into the Woolsthorpe parish registers of births and burials and found almost nothing: the information 'lost, destroyd, or obliterated; for want of care and due preservation'. The national records, he railed, were 'the most neglected! ... committed to a parish clark, illiterate, that can scarcely write, sottish, or indolent: a task on which the fortunes and emoluments of the whole kingdom in a great measure depends'. In an old town chest, a tattered vellum leaf bore this datum under the heading *baptiz'd anno 1642*: 'Isaac sonne of Isaac and Hanna Newton Jan 1'.[45]

In eighty-four years he had amassed a fortune: household furniture, much of it upholstered in crimson; crimson curtains, a crimson mohair bed and crimson cushions; a clock; a parcel of mathematical instruments and chemical glasses; several bottles of wine and cider; thirty-nine silver medals and copies in plaster of Paris; a vast library with nearly two thousand books and his many secret manuscripts; gold bars and coins – the whole estate valued at £31,821,[46] a considerable legacy.

Yet he left no will.

Notes

A word about dates. In the time with which we are concerned, the English calendar ran at first ten and then eleven days behind the calendar in most of Europe. I use the English dates. Meanwhile, the year in England was considered to start 25 March, not 1 January. So, for example, when Newton died on 20 March, they reckoned it was 1726 in England but 1727 elsewhere. From our anachronistic point of view, it was 1727, so I use the Continental – modern – years.

A word about language. Mostly I follow the spelling and style of the original texts. But where Newton (and others) compressed words to 'y^e', 'w^{ch}', 'y^t', &c., I have modernised the orthography for the sake of readability.

EPIGRAPH Newton's recollection, the year before he died, of having made the first reflecting telescope; recorded by his niece's husband, John Conduitt, memorandum, 31 August 1726, Keynes MS 130.10.
1. 'What a lesson to the vanity and presumption of philosophers!' exclaimed his first biographer, Brewster, in 1831 (*The Life of Sir Isaac Newton*, p. 303). Newton, who read incessantly and remained unsettled, was echoing Milton (*Paradise Regained*, IV, 322–30):

who reads
Incessantly, and to his reading brings not
A spirit and judgment equal or superior,
(And what he brings what needs he elsewhere seek?)
Uncertain and unsettled still remains,
Deep-versed in books and shallow in himself,
Crude or intoxicate, collecting toys
And trifles for choice matters, worth a sponge,
As children gathering pebbles on the shore.

2. Stukeley, *Memoirs*, p. 34.

3. Having compared them as lovers, Voltaire added judiciously, 'One can admire Newton for that, but must not blame Descartes.' *Letters on England*, 14, pp. 68–70.

4. Nor did he persuade us quickly. A few years before his death, a scholarly author could rail against Newton's conception of gravity ('this Cause, which looks as monstrous as any of the Fictions of Antiquity') without condescending to use the word: 'That it is a Virtue or Power which Bodies have to attract or draw one another; that every Particle of Matter has this Power or Virtue; that it reaches to all Places at all Distances, and penetrates to the Center of the Sun and Planets; that it acts not upon the Surfaces of Bodies as other Natural Agents, but upon their whole Substance or solid Content, &c. and if so, what a strange Thing must it be.' Gordon, *Remarks*, p. 6.

5. Hermann Bondi, 'Newton and the Twentieth Century – A Personal View', in Fauvel et al., *Let Newton Be!*, p. 241.

6. *Principia* (Motte), p. 6.

7. As Einstein himself knew well. Hermann Bondi commented: 'When I talk on special relativity, I always say that Einstein's contribution has a name for being difficult, but this is quite wrong. Einstein's contribution is very easy to understand, but unfortunately it rests on the theories of Galileo and Newton which are very

difficult to understand!' 'Newton and the Twentieth Century — A Personal View', in Fauvel et al., *Let Newton Be!*, p. 245.
8. *Opticks*, Foreword, p. lix.

1: WHAT IMPLOYMENT IS HE FIT FOR?

1. Barnabas Smith was sixty-three and well off; Hannah Ayscough probably about thirty; their marriage was negotiated by one of the rector's parishioners, for a fee, and by her brother. It was agreed that Isaac would remain at Woolsthorpe and that Smith would give him a parcel of land. She brought to the marriage a parcel with an income of £50.

2. One skirmish broke out near Grantham on 13 May 1643; fighting continued sporadically nearby through the summer and occasionally during the rest of the decade.

3. Cf. Clay, *Economic Expansion and Social Change*, pp. 8–9.

4. Merchants were expected to 'have knowledge and cunning in reading and writing' as well as 'the knowledge and feate of Arithmetike', if not with pen then with counters on a board. Hugh Oldcastle, *A Briefe Introduction and Maner how to keepe Bookes of Accompts* (1588), quoted in Thomas, 'Numeracy in Modern England', p. 106.

5. When he was twenty, a student at Trinity College, he suffered a sort of crisis of conscience around Whitsunday and wrote down – in a private shorthand – a catalogue of his sins. Among the early sins he included 'Threatning my father and mother Smith to burne them and the house over them' and 'Wishing death and hoping it to some'. He also recalled 'peevishness' with his mother and half-sister, striking his sister and others, 'having uncleane thoughts words and actions and dreamese', and many episodes of lying and violating the Sabbath ('Thy day'). Westfall, 'Short-Writing and the State of Newton's Conscience', p. 10.

6. Stukeley, *Memoirs*, p. 43: 'He showd another method of indulging his curiosity to find out the sun's motion, by making dyals of divers forms and constructions every where about the house, in his own chamber, in the entrys and rooms where ever the sun came.'

7. The analemma.

8. Stukeley, *Memoirs*, p. 43: 'and made a sort of almanac of these lines, knowing the day of the month by them, and the suns entry into signs, the equinoxes and solstices. So that Isaacs dyals, when the sun shined, were the common guide of the family and neighbourhood.'

9. *Henry VI*, Part 3, II.v.21.

10. Eventually he wrote:

It is indeed a matter of great difficulty to discover and effectually to distin-guish the true motions of particular bodies from the apparent, because the parts of that immovable space in which those motions are performed do by no means come under the observation of our senses. Yet the thing is not altogether desperate; for we have some arguments to guide us, partly from the apparent motions, which are the differences of the true motions; partly from the forces, which are the causes and effects of the true motions. For instance, if two globes, kept at a given distance one from the other by means of a cord that connects them, were revolved about their common center of gravity, we might, from the tension of the cord, discover the endeavour of the globes to recede from the axis of their motion, and from thence we might compute the quantity of their circular motions . . . *Principia* (Motte), p. 12.

11. Couth, ed., *Grantham during the Interregnum, 1641–1649.*

12. Stukeley, *Memoirs*, p. 43. Other, presumably Newtonian, crude diagramming has been uncovered. Whiteside ('Isaac Newton: Birth of a Mathematician', p. 56) assessed them coolly: 'It would need the blindness of maternal love to read into these sets of intersecting circles and scrawled line-figures either burgeoning artistic prowess or mathematical precocity.'

13. It was long thought that Newton had no mathematical training as a schoolboy, but Stokes's own notebook, 'Notes for the Mathematicks', exists in the Grantham Museum (D/N 2267). Whiteside, 'Newton the Mathematician', in Bechler, *Contemporary Newtonian Research*, p. III. For acres cf. Petty, *Political Arithmetick*, and John Worlidge, *Systema Agriculturæ* (London: Dorling, 1687).

14. Quoted in Manuel, *Portrait*, pp. 57–58. The 'Latin Exercise Book', originally among the papers of the Portsmouth Collection, is in private hands. Manuel adds: 'There is an astonishing absence of positive feeling. The word *love* never appears, and expressions of gladness and desire are rare. A liking for roast meat is the only strong sensuous passion.'

15. Burton, *Anatomy of Melancholy*, p. 14.

16. More fully: 'Though there were many Giants of old in Physick and Philosophy, yet I say with Didacus Stella: A dwarf standing on the shoulders of a Giant may see farther than the Giant himself; I may likely add, alter, and see farther than my predecessors.' This is neither the beginning nor the end of the story of this aphorism. For that, one must read Merton, *On the Shoulders of Giants*.

17. Burton, *Anatomy of Melancholy*, p. 423.

18. Ibid., p. 427.

19. This notebook was mentioned soon after his death by his niece's husband, John Conduitt; then it disappeared for several centuries; then it reappeared in the 1920s in the possession of the Pierpont Morgan Library, where it remains (MA 318). Cf. David Eugene Smith, 'Two Unpublished Documents of Sir Isaac Newton', in Greenstreet, *Isaac Newton*, pp. 16–34; Andrade, 'Newton's Early Notebook'; and the original Bate, *Mysteryes*.

20. Stukeley, *Memoirs*, p. 42.

21. Bate, *Mysteryes*, p. 81.

22. Dictionaries and encyclopedias ('circles' of knowledge) barely existed, but he might have seen John Withals, *A Shorte Dictionarie for*

Yonge Begynners (1556), which arranges words under subject headings; Robert Cawdry, *Table Alphabeticall Contayning and Teaching the True Writing and Understanding of Hard Usuall English Words* (1604); Francis Gregory, *Nomenclatura Brevis Anglo-Latinum.*

2: SOME PHILOSOPHICAL QUESTIONS

1. Stukeley, *Memoirs*, pp. 46—49.
2. A few years later, as a new undergraduate at Cambridge, he drew diagrams from memory that illustrate classic fluid mechanics – or rather, what would have been fluid mechanics, had this science yet been invented. He guessed to associate air and water resistance: '. . . for you may observe in water that a thing moved in it doth carry the same water behind it . . . or at least the water is moved from behind it with but a small force as you may observe by the motes in the water . . . the like must hapen in aire . . .' *Questiones*, 'Of Violent Motion', Add MS 3996, p. 21.
3. From a list of sins he set down three years later: 'Refusing to go to the close at my mothers command'; 'Punching my sister'; 'Peevishness with my mother'; 'With my sister'. Westfall, 'Short-Writing and the State of Newton's Conscience', pp. 13f.
4. Westfall, *Never at Rest*, p. 53.
5. Trinity College Note Book, MS R4.48. His tutor was Benjamin Pulleyn. He had chamber fellows but formed no friendships.
6. Notebook in the Fitzwilliam Museum, Cambridge, as transcribed by Westfall, 'Short-writing and the State of Newton's Conscience'. Westfall comments: 'We are forced to conclude either that Newton's young manhood had been remarkably pure or that his power of self-examination was remarkably under-developed. Probably we should reach both conclusions.'
7. Edward Ward, *A Step to Stir-Bitch-Fair* (London: J. How, 1700);

Daniel Defoe, *Tour through the Whole Island of Great Britain* (1724). Stourbridge Fair was the model for Vanity Fair in John Bunyan's *Pilgrim's Progress*.

8. Aristotle, *Nicomachean Ethics*, II: 1.

9. And 'becoming hot or sweet or thick or dry or white'. Aristotle, *Physics*, trans. R. P. Hardie and R. K. Gaye, VII: 2.

10. Ibid., VIII: 4.

11. Ibid., VII: 1.

12. Cf. ibid., III: 1: 'It is the fulfilment of what is potential when it is already fully real and operates not as itself but as movable, that is motion. What I mean by "as" is this: Bronze is potentially a statue.'

13. Exception: *Sidereus Nuncius*, published in Venice in 1610. Newton acquired a version of this when he was in his forties (Harrison, *The Library of Isaac Newton*, p. 147). It was first translated into English in 1880.

14. Some biographers have suggested that Newton invented this phrase, but Aristotle expresses the sentiment in *Nicomachean Ethics* I: 6, and the Latin motto is attributed to him in Diogenes Laërtius, *De vitis dogmatibus et apophtegmatibus clarorum philosophorum*, a copy of which Newton owned. For more exhaustive detective work on the slogan see Guerlac, 'Amicus Plato and Other Friends', in *Newton on the Continent*.

As he wrote, Newton was reading closely – and sometimes disputing – Walter Charleton (*Physiologia Epicuro-Gassendo-Charltoniana*), Descartes (a partial collected works, in Latin), the Platonist Henry More (*The Immortality of the Soul*) and the contemporary experimenter Robert Boyle. The definitive analysis of the *Questiones*, including a careful transcription, is McGuire and Tamny, *Certain Philosophical Questions*.

The notebook is in the Cambridge University Libraries as Add MS 3996. My citations use Newton's page numbers.

15. *Questiones*, p. 1.

16. Ibid., p. 6.

17. Ibid., p. 32.

18. Ibid., p. 21.

19. Ibid., p. 19.

20. 'Siccity': dryness.

21. Coastal-dwelling people in every part of the world had noticed coincidences in timing between the flow of tides and the changing of the moon, as well as the sun. Near shores and harbours of the North Atlantic, in particular, monks had been saving data – though not disseminating it – for hundreds of years.

3: TO RESOLVE PROBLEMS BY MOTION

1. Conduitt, 'Memorandum relating to Sr Isaac Newton given me by Mr Demoivre in Novr 1727':

In 63 being at Stourbridge fair bought a book of Astrology ... Read in it till he came to a figure of the heavens which he could not understand for want of being acquainted with Trigonometry. Bought a book of Trigonometry, but was not able to understand the Demonstrations. Got Euclid to fit himself for understanding the ground of Trigonometry. Read only the titles of the propositions, which he found so easy to understand that he wondered how any body would amuse themselves to write any demonstrations of them. Began to change his mind when he read that Parallelograms upon the same base & between the same Parallels are equal, & that other proposition that in a right angled Triangle the square of the Hypothenuse is equal to the squares of the other two sides.

Cf. Keynes MS 130.4; and *Math* I: 15.

2. Thus Whiteside: 'We are, too, perhaps a little disappointed that

Newton read so little of standard contemporary mathematical works, or if he did has left no hint – nowhere in his early autograph papers do we find the names of Napier, Briggs, Desargue, Fermat, Pascal, Kepler, Torricelli, or even Archimedes and Barrow.' 'Sources and Strengths of Newton's Early Mathematical Thought', in Palter, *Annus Mirabilis*, p. 75. Apart from Newton's notes, his second- and third-hand recollections of his reading, including the 'book of Astrology', survive in an account by Abraham DeMoivre (Add MS 4007); also *Corres* VII: 394.

3. Some survived infection, but not many. In Cambridge the final 'Plague Bill' reported a total of 758 deaths from 5 June to 1 January, all but nine from the plague. About half that number were infected and recovered. Leedham-Green, *Concise History*, p. 74.

4. This is the only surviving letter either to or from Newton's mother (or, for that matter, any close relative). The edges are torn and some words are missing. *Corres* I: 2.

5. Add MS 4004.

6. The 'year' – traditionally seen by Newtonians as the *annus mirabilis* – covered eighteen or twenty or twenty five months. Sophisticated Newtonians sometimes prefer to speak of the 'myth' of the *annus mirabilis*. For example, Derek Gjertsen debunks the myth sternly: 'The description is clearly misleading, for . . . no special priority can be given to either 1665 or 1666 . . . It remains true, none the less, and without too much exaggeration, that in a remarkably short period the twenty-four-year-old student created modern mathematics, mechanics, and optics. There is nothing remotely like it in the history of thought.' Gjertsen, *Newton Handbook*, p. 24. Cf. Whiteside ('Newton the Mathematician', in Bechler, *Contemporary Newtonian Research*, p. 115): 'Never did seventeenth-century man build up so great a store of mathematical expertise, much of his own discovery, in so short a time.'

Anyway, Newton's stay in Woolsthorpe extended over about

twenty months, broken by a temporary return to Cambridge in the spring of 1666.

7. Alfred North Whitehead noted that Europe knew less mathematics in 1500 than Greece knew in the time of Archimedes. Davis and Hersh, *Mathematical Experience*, p. 18.

8. 'Thrice happy he, who, not mistook,/Hath read in nature's mystic book!' Andrew Marvell, 'Upon Appleton House, to My Lord Fairfax'.

9. Galileo, *Il Saggiatore* (1623), in *The Controversy on the Comets of 1618*, pp. 183–84.

10. Elliott, 'Isaac Newton's "Of an Universall Language",' p. 7.

11. Whiteside, 'Newton the Mathematician', in Bechler, *Contemporary Newtonian Research*, pp. 112–13. Newton's annotated student copy of the *Elements*, Trinity College Library, NQ.16.201.

12. John Conduitt's romanticised account (Keynes MS 130.4, in *Math* I: 15–19):

He then young as he was took in hand Des-Cartes's Geometry (that book which Descartes in his Epistles with a sort of defiance says is so difficult to understand). He began with the most crabbed studies & books, like a high spirited horse who must be first broke in crabbed grounds & the roughest & steepest ways, or could otherwise be kept within no bounds. When he had read two or three pages & could understand no farther he being too reserved and modest to trouble any person to instruct him begain again & got over three or four more till he came to another difficult place, & then began again & advanced farther & continued so doing till he not only made himself master of the whole without having the least light or instruction from any body, but discovered the errors of Descartes ...

He read it in Schooten's Latin translation in the summer of 1664. Newton's own reminiscences of his mathematical development tended to minimise the role of Descartes, but Whiteside's scholarship is conclusive: that 'the thick wad of Newton's research papers surviving from the later months of 1664 stand firm witness that it was indeed from the hundred or so pages of the *Géométrie* that his mathematical spirit took fire ... Above all, I would assert, the *Géométrie* gave him his first true vision of the universalizing power of the algebraic free variable, of its capacity to generalize the particular and lay bare its inner structure.' 'Newton the Mathematician', in Bechler, *Contemporary Newtonian Research*, p. 114.

But he also filled it with critical marginalia; e.g. '*Error, Error, non est Geom*' and '*Imperf*'. Trinity College Library, NQ.16.203.

13. 'It seems to be nothing other than that art which they call by the barbarous name of "algebra", if only it could be disentangled from the multiple numbers and inexplicable figures that overwhelm it ...' Descartes, *Regulæ ad directionem ingenii*, Regula IV: 5.

14. This new-found truth had to be stated explicitly. Mahoney ('The Beginnings of Algebraic Thought') quotes Descartes: 'Those things that do not require the present attention of the mind, but which are necessary to the conclusion, it is better to designate by the briefest symbols than by whole figures: in this way the memory cannot fail, nor will thought in the meantime be distracted by these things which are to be retained.'

15. Keynes MS 130(7), quoted by Christianson, *In the Presence of the Creator*, p. 66.

16. *Biographia Britannica* (London, 1760), V: 3241; quoted by Westfall, *Never at Rest*, p. 174.

17. The recognition of infinite series had begun with algebraic attempts to express pi; Newton's immediate predecessors, James Gregory and especially John Wallis, were the first to develop their possibilities. In the simplest sense, infinite series are implied immediately by decimal notation; in one of his earliest jotted fragments Newton wrote: 'if the fraction 10/3 bee reduced to decimall it will be 3,33333333 &c infinitely. & what doth every figure signifie but a pte of

the fraction 10/3 which therefore is divisible into infinite pts.'
Questiones, p. 65.

18. *Math* I: 134–41; Westfall, *Never at Rest*, pp. 119–21. This was, he saw, another problem in disguise, the calculation of a logarithm. Years later he recalled: 'I am ashamed to tell to how many places I carried these computations, having no other business at that time: for then I took really too much delight in these inventions.' Newton to Oldenburg, 24 October 1676, *Corres* II: 188.

19. Descartes, *Principles of Philosophy*, in *Philosophical Writings*, I: 201.

20. Even seventy years later, one of the first post-Newtonian calculus texts, John Colson's 1737 *Method of Fluxions and Infinite Series*, broached the dangerous and unfamiliar topic this way: '. . . that quantity is *infinitely divisible*, or that it may (*mentally* at least) *so far* continually diminish, as at last, before it is *totally* extinguished, to arrive at quantities which may be called *vanishing* quantities, or which are *infinitely little*, and less than any *assignable* quantity . . .' In Cohen and Westfall, *Newton: Texts*, p. 400.

21. 'Of Quantity', *Questiones*, p. 5; *Math* I: 89.

22. *Questiones*; cf. *Math* I: 90, n. 8.

23. Galileo, *Discorsi*.

24. *Math* I: 280.

25. Ibid., 282.

26. Ibid., 302 and 305.

27. *Questiones*, p. 10.

28. *Questiones*, p. 68.

29. Cf. *Math* I: 377; Michael Mahoney, 'The Mathematical Realm of Nature', in Garber and Ayers, *Cambridge History of Seventeenth-Century Philosophy*, p. 725.

30. *Math* I: 29.

31. 'To find the velocitys of bodys by the lines they describe.' *Math* I: 382.

32. *Math* I: 273.

33. Much later he recalled: 'When I am investigating a truth or the solution to a Probleme I use all sorts of approximations and neglect to write down the letter o, but when I am demonstrating a Proposition I always write down the letter o & proceed exactly by the rules of Geometry.' Add MS 3968.41.

34. *Math* I: 377ff., I: 392ff. and I: 400ff. The tract of October 1666 (Add MS 3958) was published for the first time 296 years later in Hall and Hall, *Unpublished Scientific Papers*, pp. 15–65.

35. *Math* I: 402.

36. As Koyré puts it, 'To have achieved this transformation is the undying merit of Newton ... Mathematical entities have to be, in some sense, brought nearer to physics, subjected to motion, and viewed not in their "being" but in their "becoming" or in their "flux".' *Newtonian Studies*, p. 8.

4: TWO GREAT ORBS

1. The last authoritative twentieth-century account of the Scientific Revolution, Steven Shapin's *Scientific Revolution*, began, 'There was no such thing as the Scientific Revolution, and this is a book about it.'

2. Goodstein and Goodstein, *Lost Lecture*, p. 39.

3. 'The appearance of Newton's *Principia* in 1687 changed all this ... [It] made continued support for Aristotle's geocentric cosmology untenable. After 1687, medieval cosmology became irrelevant, because it no longer represented even a minimally plausible alternative to Newtonian cosmology. Unlamented, it simply faded away.' Grant, *Planets, Stars, and Orbs*, p. 10.

4. Yet I. Bernard Cohen sees the Copernican revolution as 'a fanciful invention of eighteenth-century historians'. The revolution, Cohen asserts, 'was not at all Copernican, but was at best Galilean and

Keplerian'. *Revolution in Science*, p. *x*. Meanwhile, Cohen and other scholars suggest that Newton's reading, wide-sweeping though it became, may have never included Galileo's *Discorsi* or anything of Kepler. Nor, at his death, did his considerable library contain any work by Ptolemy, Copernicus or Tycho Brahe. Cf. Whiteside in *Math*, VI: 3 n. and 6 n.

5. Now we say these were the first two of Kepler's three 'laws'. We conventionally date these to 1609, when he published his great work, *Astronomia Nova*. He put forth a notion of gravity, too: 'Gravity is the mutual tendency of cognate bodies to join each other (of which kind the magnetic force is).' Nevertheless, by the time of the *Principia*, at the far end of the century, few astronomers accepted Kepler's ideas as firm truths; nor did Newton, in the *Principia*, see Kepler as a significant precursor. 'It seems clear,' I. B. Cohen remarked, 'that there was no Keplerian revolution in science before 1687.' *Revolution in Science*, p. 132; Whiteside, 'Newton's Early Thoughts on Planetary Motion', p. 121; Gjertsen, 'Newton's Success', in Fauvel et al., *Let Newton Be!*, p. 25.

6. Galileo, *The Starry Messenger*, in *Discoveries and Opinions*, pp. 27f.

7. The only mathematics, except that Galileo declared the moon's distance to be sixty diameters of the earth – off by a factor of two – and made a brief computation of the height of lunar mountains, declaring (correctly) that they were as high as four miles and (incorrectly) that the earth's mountains never reached as high as one mile. For a moment, it was easier to see the moon than the earth.

8. Two years later: *Discourse concerning a New Planet; tending to prove, that it is probable our Earth is one of the Planets*. Wilkins also wrote another book cherished by the young Newton, *Mathematical Magick*.

9. Wilkins, *Mathematical and Philosophical Works*, pp. 34 and 11.

10. Bacon, 'Of Tribute: Praise of Knowledge', *Works* VIII: 125.

11. Bacon, *Novum Organum*, pp. 217 and 260.

12. Wilkins, *Mathematical and Philosophical Works*, pp. 47, 49, 97, 100,

110–13. For flying to the moon, Wilkins did wonder about the cargo problem: 'Nor can we well conceive how a man should be able to carry so much luggage with him, as might serve for his *viaticum* in so tedious a journey.'

13. Ibid., pp. 4 and 13: 'For it is probable that the earth of that other world would fall down to this centre, and so mutually the air and fire here ascend to those regions in the other; which must needs ... cause a great disorder ...'

14. Ibid., pp. 61 and 14.

15. Ibid., p. 114.

16. He copied bits of Wilkins into his Grantham notebook (cf. Manuel, *Portrait*, p. 11, and Gjertsen, *Newton Handbook*, p. 612). Wilkins also expounded systems of 'secret writing' – how to hide one's meaning through obscure or invented or encoded characters (*Mercury; or, the Secret Messenger*, 1641). He became a doctor of divinity and a prominent Parliamentarian, married Oliver Cromwell's sister, and soon after was made Master of Trinity College, a preferment he held only briefly before being ousted upon the restoration of Charles II. He moved to London and became a council member of the new Royal Society.

17. Herivel, *Background to Newton's Principia*, p. 67; Add MS 3968.41; Westfall, *Never at Rest*, p. 143.

18. The river flows from four memoirists in particular: his niece, Catherine Barton; Marton Folkes, vice-president of the Royal Society; Barton's husband, John Conduitt; and Newton's first would-be biographer, William Stukeley. 'The notion of gravitation came into his mind,' Stukeley wrote (*Memoirs*, p. 20), '... occasion'd by the fall of an apple, as he sat in a contemplative mood.'

Voltaire related the story first in *An Essay on Epick Poetry* and then in *Letters on England* (p. 75): 'Having retired to the country near Cambridge in 1666, he was walking in his garden, saw some fruit falling from a tree, and let himself drift into a profound meditation on this weight, the cause of which all the scientists have vainly sought for so long and about which ordinary people never even suspect there is any mystery.'

And Conduitt: 'Whilst he was musing in a garden it came into his thought that the power of gravity (which brought an apple from the tree to the ground) was not limited to a certain distance from the earth but that this power must extend much farther than was usually thought. Why not as high as the moon said he to himself . . .' Keynes MS 130.4.

The apple story took on an independent life and evolved over centuries. Perhaps its most wonderful feature is how often, by the twentieth century, the apple was supposed to have struck Newton on the head. This may not have been necessary.

Westfall argues, appealingly (*Never at Rest*, p. 155): 'The story vulgarizes universal gravitation by treating it as a bright idea.' Of course! Yet it was a bright idea. We feel this deeply. Surely that's why the story has so rooted itself in our collective consciousness. The bright idea was a crystallisation of a pre-existing unconscious knowledge – shared by animals and children – that objects fall to the ground. The bright idea was that this behaviour implied a force – to be named and then studied and measured. 'A bright idea cannot shape a scientific tradition,' Westfall adds, and this, too, seems self-evident. But it did.

19. Galileo, *Two New Sciences*, p. 166, quoted in Cohen, *Franklin and Newton*, p. 103.

20. One detailed set of calculations fills the so-called Vellum Manuscript – the reverse side of a lease. Add MS 3958.45; Herivel, *Background to Newton's Principia*, pp. 183–191.

21. Where the 'cubit' is the distance from elbow to fingertip. Herivel, *Background to Newton's Principia*, p. 184.

22. Thomas Salusbury, 1665.

23. Herivel, *Background to Newton's Principia*, p. 186.

24. 'The cubes of their distances from the Sun are reciprocally as the squares of the numbers of revolutions in a given time: the endeavours of receding from the Sun will be reciprocally as the squares of the distances from the Sun.' Add MS 3958, in Herivel, *Background to Newton's Principia*, p. 197; Westfall, *Never at Rest*, p. 152. In the same spirit: *Principia*, Book III, Proposition 10, Corollary 3 and Corollary 5 (first edition), where Newton explicitly considers the sun's heating of the planets as a function of distance.

25. This was eventually known as Kepler's third law, the law of periods.

26. Herivel, *Background to Newton's Principia*, p. 141. Descartes had proposed such a principle for bodies both in motion and at rest, though not for circular motion. It still defied people's intuition about moving objects. 'That when a thing lies still, unless somewhat else stir it, it will lie still for ever, is a truth that no man doubts of,' Hobbes wrote in 1651. 'But that when a thing is in motion, it will eternally be in motion, unless somewhat else stay it, though the reason be the same (namely, that nothing can change itself), is not so easily assented to.' People get tired and stop moving, so they imagine inanimate objects do, too. 'From hence it is that the schools say, heavy bodies fall downwards out of an appetite to rest, and to conserve their nature in that place which is most proper for them.' *Leviathan*, II.

27. Herivel, *Background to Newton's Principia*, p. 158.

28. Ibid., p. 153.

29. Nor was Latin any better. In trying systematically to define concepts in terms of simpler or more basic concepts, he always reached a wall – a problem of infinite regress. Yet he kept trying. In an undated notebook (Add MS 4003): 'The terms *quantity, duration* and *space* are too well known to be susceptible of definition by other words.

Def. 1. Place [*Locus*] is a part of space which something fills evenly.

Def. 2. Body [*Corpus*] is that which fills place.

Def. 3. Rest [*Quies*] is remaining in the same place.

Def. 4. Motion [*Motus*] is change of place.'

(In Hall and Hall, *Unpublished Scientific Papers*, pp. 91 and 122.)

30. Herivel, *Background to Newton's Principia*, p. 155.

5: BODYS & SENSES

1. Add. MS 3996.

2. Andrew Marvell, 'A Dialogue Between the Soul and Body'.

3. 'Immagination & Phantasie & invention', *Questiones*.

4. Add MS 3975.

5. *Questiones*, p. 43.

6. Newton to Locke, 30 June 1691, *Corres* III: 365.

7. Hooke, *Micrographia*, preface.

8. Letter of John Wallis, quoted in Charles Richard Weld, *History of the Royal Society*, I: 30; Ornstein, *Role of Scientific Societies*, pp. 93 and 95; *Phil. Trans.* 1 (March 1665). Several such societies, on a regional scale, had been formed in Naples and Florence; the next national scientific society, the *Académie des Sciences*, was founded in Paris four years later.

9. Wallis letter, in Weld, *History of the Royal Society*, I: 30; Ornstein, *Role of Scientific Societies*, p. 95.

10. Horace, *Epistles* I: 1, 14: 'Nullius addictus iurare in verba magistri . . .'

11. Bacon, *Novum Organum*, p. 169.

12. 'An Account of a Dog dissected by Mr. Hook', in Sprat, *History of the Royal Society*, p. 232; 'Espinasse, *Robert Hooke*, p. 52.

13. Pepys, *Diary*, 30 May 1667. 'fine experiments . . . of colours, loadstones, microscopes, and of liquors . . . among others, of one that did while she was there turn a piece of roasted mutton into pure blood, which was very rare . . . After they had shown her many experiments, and she cried out still she was full of admiration, she departed . . .'

14. Hooke tracked his internal weather with equal diligence. A typical journal entry: 'Slept a second sleep, sweat and ✳ [ejaculation]. Rose at 11. Eat broth drank port. Belly loosned. Much refresht. 2 stools. DH. With Aubery. Haak chesse. To Garaways. With Tompion and Sir J. Mores. at 7 till 9. Belly loose. Smell well amended. Smokd 4 pipes. Chocolat H. 1. Port. Slept. Sweat.'

15. Hooke, *Micrographia*, preface.

16. Ibid., p. 3.

17. 'This Experiment therefore will prove such a one as our thrice excellent Verulam calls *Experimentum Crucis*, serving as Guide or Land-mark, by which to direct our course in the search after the true cause of Colours. Affording us this particular negative Information, that for the production of Colours there is not necessary either a great refraction, as in the Prisme; nor Secondly, a determination of Light and shadow, such as is both in the Prisme and Glass-ball.' Ibid., p. 54.

18. A 'pellucid body', as Hooke put it, 'where there is properly no such refraction as Des Cartes supposes his Globules to acquire a vorticity by'. Ibid.

19. Ibid., p. 64.

20. Ibid., p. 55. He did not care to admit what he did not know. 'It is not my business in this place to set down the reasons why this or that body should impede the Rays more, others less: as why Water should transmit the Rays more easily, though more weakly than air.'

21. Ibid., p. 67.

22. Newton's notes: 'Out of Mr Hooks Micrographia'. Add MS 3958(3).1.

23. Bacon, *Novum Organum*, p. 30.

6: THE ODDEST IF NOT THE MOST
CONSIDERABLE DETECTION

1. Westfall, *Never at Rest*, p. 179.
2. 1669 purchases in the Fitzwilliam notebook.
3. *Math* II: 99–150; W. W. Rouse Ball, 'On Newton's Classification of Cubic Curves', *Proceedings of the London Mathematical Society*, 22 (1890–91): 104–43.
4. Barrow catered to Newton's skittishness by telling Collins: 'I pray having perused them so much as you thinke good, remand them to me; according to his desire, when I asked him the liberty to impart them to you. And I pray give me notice of your receiving them with your soonest convenience; that I may be satisfyed of their reception; because I am afraid of them; venturing them by the post.' (31 July 1669, *Corres* I: 6.) Newton did eventually permit the publication of *De Analysi per Æquationes Infinitas* – in 1711, when he was sixty-nine.
5. Barrow to Collins, 20 August 1669, *Corres* I: 7.
6. Newton to Collins, January 1670, *Corres* I: 9.
7. Newton to Collins, February 1670, *Corres* I: 12.
8. Gregory to Collins, September 1670, 5, *Corres* I: 18.
9. *Lectiones opticæ & geometricæ: in quibus phænomenon opticorum genuinæ rationes investigantur, ac exponuntur: et generalia curvarum linearum symptomata declarantu* (London, 1674). Scholars have debated Newton's reticence with Barrow. I. Bernard Cohen found it inconceivable that Newton could have withheld his knowledge from Barrow at this crucial point; he speculated that Barrow just lacked the time or inclination to start his optical work anew (*Franklin and Newton*, p. 52). But, plausibly, Christianson saw 'a *prima facie* case of deceit on Newton's part, a hypocritical laughing up his sleeve at the work of a man who was about to advance his career' (*In the Presence of the Creator*, p. 125).

10. *Lectiones*, p. 108, quoted in Shapiro, *Optical Papers*, I: 15 n.

11. Barrow was appointed Royal Chaplain and then, three years later, Master of Trinity College.

12. *Math* III: xx.

13. 'So few went to hear Him, & fewer that understood him, that oftimes he did in a manner, for want of Hearers, read to the walls . . . usually staid about half an hour, when he had no Auditors he commonly return'd in a 4th part of that time or less.' Humphrey Newton, quoted by Conduitt, Keynes MS 135; in *Math* VI: xii n. The historical record contains not a single recollection from anyone who heard Newton lecture.

14. Shapiro, *Optical Papers* I: 47. This first lecture was delivered in January 1670 and a version deposited in the library, belatedly, in 1674.

15. 'I left off my aforesaid Glass-works; for I saw, that the perfection of Telescopes was hitherto limited, not so much for want of glasses truly figured according to the prescriptions of Optick Authors, (which all men have hitherto imagined), as because that Light it self is a *Heterogeneous mixture of differently refrangible rays*.' Newton to Oldenburg, 6 February 1672, *Corres* I: 40.

16. The original idea of a reflecting telescope seems to have been James Gregory's, though Gregory never succeeded in building one. *Corres* I: 159.

17. *Corres* I: 3.

18. Sprat, *History of the Royal Society*, p. 20.

19. Indeed, in 1664 they appointed a committee for improving the English language. It never produced anything definite. (Lyons, *Royal Society*, p. 55.)

20. Hobbes, *Leviathan*, V.

21. Galileo to Mark Welser, 4 May 1612, trans. Stillman Drake, in *Discoveries and Opinions of Galileo*, p. 92.

22. E.g. *Corres* I: 35.

23. Samuel Sorbière, *A Voyage to England* (1709), quoted in Hall, *Henry Oldenburg*, p. 52.

24. *Transactions* was a plausible word for this new creature, the serial publication, though the word did not stick. The terms *journal* and *periodical* did not yet exist in this context. Words like *gazette*, *pamphlet* and *tract* had unpleasant connotations, as Adrian Johns notes ('Miscellaneous Methods', p. 162).

The *Philosophical Transactions* stands as the first scientific journal, almost. Derek Gjertsen notes that the Academia del Cimento began printing its proceedings in 1657 and continued for about ten years, and that the *Journal des Sçavans* began appearing in Paris two months before the *Philosophical Transactions* but encompassed history and law as well as natural philosophy. *Newton Handbook*, p. 431. About three hundred copies of the first issue were sold. The journal never came close to bringing Oldenburg the profit he hoped for.

25. *Phil. Trans.* 3: 632; 3: 693.

26. John Evelyn, *Diary*, III: 288–89, 295 and 325.

27. Samuel Butler, 'The Elephant in the Moon' (1759).

28. *Phil. Trans.* 1: 10; 3: 792; 3: 704; 3: 43; 3: 115.

29. Notes 'Out of the Hystory of the Royall Society', Add MS 3958c.

30. Oldenburg to Newton, 2 January 1672, *Corres* I: 29, and I: 3.

31. The telescope, or 'perspectives', did not make a deep impression on all assembled. John Evelyn, later famous for his diaries, recorded the event this way: 'To the R. Society; where were produced new invented Perspectives, a letter from Grene-land, of recovering men that had ben drown'd, we had also presented from Iseland some of the Lapis Obsidialis.' *Diary of John Evelyn*, III: 601.

32. Newton to Oldenburg, 6 January 1672, *Corres* I: 33.

33. Newton to Oldenburg, 18 January 1672, *Corres* I: 35.

7: RELUCTANCY AND REACTION

1. G. N. Watson, 'Trinity College in the Time of Newton', in Greenstreet, *Isaac Newton*, p. 146.

2. Newton to Oldenburg, 6 February 1672, *Corres* I: 40. This is a correct account of the Magnus effect, named after Heinrich Gustav Magnus, who 'discovered' it in 1852, 180 years after Newton.

3. *Phil. Trans.* 80 (February 1672): 3075.

4. Newton to Oldenburg, 6 February 1672, *Corres* I: 40.

5. Thomas Kuhn lists Seneca (first century), Witelo (thirteenth century), Descartes, Marcus, Boyle and Grimaldi, as well as Hooke, among those who had seen 'the celebrated phenomena of colors'. 'Newton's Optical Papers', in Cohen, *Papers and Letters*, p. 29. Much scholarship considers the question of when and where Newton obtained his prisms and, for that matter, when and where he first conducted this experiment. Various pieces of evidence, including this letter, the Fitzwilliam Notebook and the recollections of Conduitt fifty years later, contradict one another.

6. *Instantia Crucis*, crucial instance.

7. *Questiones*, p. 69.

8. *Phil. Trans.* 80 (February 1672): 3083.

9. For that matter, the letter was the first major scientific work published in a journal.

10. Newton to Oldenburg, 6 February 1672, *Corres* I: 40, pp. 96–97 and n. 19.

11. And: 'How doth the formost weake pulse keepe pace with the following stronger?' Add MS 3958(3).1, notes 'Out of Mr Hooks Micrographia'.

12. *Phil. Trans.* 80 (February 1672): 3085.

13. As Kuhn notes: 'To destroy the modification theory it was necessary to notice a *quantitative* discrepancy between the elongation

predicted by that theory and the elongation actually observed, and this required an experimenter with a knowledge of the mathematical law governing refraction (not announced until 1637) and with considerable experience in applying the law to optical problems. In 1666 these qualifications were uniquely Newton's.' 'Newton's Optical Papers', in Cohen, ed., *Papers and Letters*, p. 32.

14. Casper Hakfoort, 'Newton's Optics: The Changing Spectrum of Science', in Fauvel et al., p. 84.

15. E.g. *Corres* I: 41.

16. Newton to Oldenburg, 24 October 1676, *Corres* II: 188.

17. Hooke to Oldenburg, 15 February 1672, *Corres* I: 44. Newton retorted that Hooke might as well speak of the 'light in a piece of wood before it be set on fire'. Newton to Oldenburg, 11 June 1672, *Corres* I: 67.

18. Pardies to Oldenburg, 30 March 1672, *Corres* I: 52.

19. Newton to Oldenburg, 13 April 1672, *Corres* I: 55. Pardies replied politely that Newton had answered some of his objections and that *hypothesis* had merely been the first word that came to mind.

20. He continued: 'I shall now take a view of Mr Hooks Considerations on my Theories. And those consist in ascribing an Hypothesis to me which is not mine . . . & in denying some thing the truth of which would have appeared by an experimentall examination.' Newton to Oldenburg, 11 June 1672, *Corres* I: 67.

21. *Corres* I: 99 and 103.

22. Newton to Oldenburg, 8 March 1673, *Corres* I: 101; Newton to Collins, 20 May 1673, *Corres* I: 110. Oldenburg to Newton, 4 June 1673, *Corres* I: 112.

23. '. . . or rather that you will favour me in my determination by preventing so far as you can conveniently any objections or other philosophicall letters that may concern me.' Newton to Oldenburg, 23 June 1673, *Corres* I: 116.

24. Newton's silence lasted from June 1673 to November 1675 – broken only by one more curt rejection: 'I have long since determined to concern my self no further about the promotion of Philosophy. And for the same reason I must desire to be excused from ingaging to exhibit yearly philosophic discourses . . . If it were my lot to be in London for sometime, I might possible take occasion to supply a vacant week or two with something by me, but that's not worth mentioning.' Newton to Oldenburg, 5 December 1674, *Corres* I: 129.

25. '*umbram captando eatinus perdideram quietam meam* . . .' Newton to Oldenburg, 24 October 1676, *Corres* II: 188.

8: IN THE MIDST OF A WHIRLWIND

1. Boyle, *The Sceptical Chymist*, p. 57. Yet he did not quite believe that gold was an element, in the modern sense.

2. Ibid., p. 3.

3. The various alternative versions of the Hypothesis are best seen in the *Correspondence*: Newton to Oldenburg, 7 December 1675, *Corres* I: 146.

4. 'to avoid circumlocation', ibid.

5. It included, besides the 'Hypothesis' (not published during his lifetime), the 'Note on the Discourse of Observations' (adapted almost intact, decades later, as Book II of the *Opticks*).

6. Overoptimistic by a factor of a thousand or so. *Corres* I: 391 n.; Birch, *History of the Royal Society*, III: 303; S. I. Vavilov, 'Newton and the Atomic Theory', in Royal Society, *Newton Tercentenary Celebrations*, p. 48.

7. *Corres* I: 146.

8. *Corres* I: 366.

9. Newton's physical intuition failed him here, in that he neglected another source of damping for a pendulum in vacuum – friction within the cord – but years later, soon before the *Principia*, he

repeated this experiment more carefully and began to lose faith in the ether. Cf. Westfall, 'Uneasily Fitful Reflections on Fits of Easy Transmission', in Palter, *Annus Mirabilis*, pp. 93 and 100 n.; also 'De Ære et Æthere', Add MS 3970.

10. *Corres* I: 368.

11. 'And they that will,' he added, 'may also suppose, that this Spirit affords or carryes with it thither the solary fewell & materiall Principle of Light; And that the vast æthereall Spaces between us, & the stars are for a sufficient repository for this food of the Sunn & Planets.' *Corres* I: 366.

12. *Physico-mathesis de lvmine, coloribvs et iride* (1665).

13. Birch, *History of the Royal Society*, III: 269; *Corres* I: 407 n.

14. Newton to Oldenburg, 21 December 1675, *Corres* I: 150.

15. Hooke and Oldenburg were at war over another matter, Oldenburg's promotion of Huygens's invention of a spiral-spring-regulated watch – previously invented, according to Hooke, by Hooke. Hooke's extant diary scarcely mentions Newton, ever, but Oldenburg is everpresent: e.g. 'the Lying Dog Oldenburg'; 'Oldenburg treacherous and a villain'. Hooke, Diary, 8 November 1675 and 28 January 1673; 'Espinasse, *Robert Hooke*, pp. 9 and 65.

16. 'These to my much esteemed friend, Mr Isaack Newton, at his chambers in Trinity College . . .' Hooke to Newton, 20 January 1676, *Corres* I: 152.

17. Newton to Hooke, 5 February 1676, *Corres* I: 154.

18. Some commentators have been pleased to note that, in literal terms, Hooke was no giant; his physique was diminutive and twisted. His contemporary John Aubrey described him in *Brief Lives* as 'but of midling stature, something crooked, pale faced, and his face but little belowe, but his head is lardge'. This hardly seems relevant to Newton's choice of trope. It is clear that *the shoulders of giants* had already lived for some centuries as a conventional expression; Robert Merton has traced its course most magisterially.

9: ALL THINGS ARE CORRUPTIBLE

1. An 'oven mouthed chimney'. Yehuda MS 34, quoted in Westfall, *Never at Rest*, p. 253 n.

2. Stukeley, *Memoirs*, pp. 60–61; Humphrey Newton's recollection, Keynes MS 135; John Wickins, Keynes MS 137.

3. Analysis of four surviving locks of Newton's hair in 1979 found toxic levels of mercury. Johnson and Wolbarsht, 'Mercury Poisoning: A Probable Cause of Isaac Newton's Physical and Mental Ills'; Spargo and Pounds, 'Newton's "Derangement of the Intellect"'. But the severity remains in doubt, as do suggestions that mercury poisoning contributed to Newton's mental troubles. See also Ditchburn, 'Newton's Illness of 1692–3'.

4. Gaule, *Pys-mantia*, p. 360.

5. Keynes MS 33. Maybe Mr. F. was Ezekial Foxcroft (Dobbs, *Foundations of Newton's Alchemy*, p. 112); at any rate the mystery, and the peopling of his papers with unidentified gentlemen, is a continual source of frustration for his biographers. 'This is only speculation, of course,' Westfall remarks, typically. 'It is not speculation that Newton had alchemical manuscripts which he must have received from someone since they did not, I believe, materialize out of thin air.' Westfall, *Never at Rest*, p. 290.

6. In the 1680s he had an amanuensis, Humphrey Newton (no relation), who recalled: 'Especially at the Spring and Fall of the Leaf, at which Times he used to imploy about 6 weeks in his Elaboratory, the Fire scarcely going out either Night or Day, he siting up one Night, as I did another, till he had finished his Chymical experiments . . . What his Aim might be, I was not able to penetrate into, but his Pains . . . made me think, he aimed at something beyond the Reach of humane Art and Industry.' Cohen and Westfall, *Newton: Texts*, p. 300.

7. *The Works of Geber Englished by Richard Russell* (reprinted London: Dent, 1928), p. 98.

8. Cinnabar was red mercuric sulphide, also known to painters as vermilion. Alchemists knew that it was a 'sublimation' of quicksilver (mercury) and brimstone (sulphur). Meanwhile, the identification of quicksilver with mercury was not perfect; alchemists also spoke of a 'philosophic mercury', a more general substance, which might be extracted from other metals as well.

9. White, *Medieval Technology*, p. 131.

10. The symbol was a pair of serpents – one male and one female – entwined about a staff.

11. Add MS 3973, quoted in Westfall, *Never at Rest*, p. 537.

12. Keynes MS 55, quoted in Dobbs, *Foundations*, p. 145.

13. *Phil. Trans.* 10:515–33.

14. 'In my simple judgment the noble Author since he has thought fit to reveale himself so far does prudently in being reserved in the rest.' Newton to Oldenburg, 26 April 1676, *Corres* II: 157. Newton concludes with regret for his unusual loquacity: 'I have been so free as to shoot my bolt: but pray keep this letter private to your self.'

15. Peter Spargo, 'Newton's Chemical Experiments', in Theerman and Seeff, *Action and Reaction*, p. 132: 'To the best of my knowledge no contemporary chemist, including Boyle, approached this degree of quantification in chemistry – nor indeed was anyone to do so until some time later.'

16. 'On Natures Obvious Laws and Processes in Vegetation', in Cohen and Westfall, *Newton: Texts*, pp. 301, 305, and 303.

17. Keynes MS 56, quoted in Westfall, *Never at Rest*, p. 299.

18. Cohen, *Revolution in Science*, p. 59.

19. 'De Gravitatione et æquipondio fluidorum', in Hall and Hall, *Unpublished Scientific Papers*, p. 151. 'I suppose that the parts of hard bodies do not merely touch each other and remain at relative rest, but that they do so besides so strongly and firmly cohere, and are so bound together, as it were by glue . . .'

20. 'And what certainty can there be in Philosophy which consists in as many Hypotheses as there are Phænomena to be explained.' Add MS 3970.3, quoted in Hutchison, 'What Happened to Occult Qualities in the Scientific Revolution?'

21. Newton to Oldenburg, 7 December 1675, *Corres* I: 146.

10: HERESY, BLASPHEMY, IDOLATRY

1. Westfall, *Never at Rest*, pp. 311–12. The 'theological notebook' is Keynes MS 2, one of those marked (by Thomas Pellett) after Newton's death 'Not fit to be printed' and then stored, unread, until Keynes acquired it in 1936.

2. He told Oldenburg and reminded him in January 1675: 'the time draws near that I am to part with my fellowship . . .' *Corres* VII: X.132.

3. From a memorandum by David Gregory, in Cohen and Westfall, *Newton: Texts*, p. 329.

4. 'The father is immoveable no place being capable of becoming emptier or fuller of him then it is by the eternal necessity of nature: all other being are moveable from place to place.' 'A Short Schem of the True Religion', Keynes MS 7, in Cohen and Westfall, *Newton: Texts*, p. 348.

5. *Principia* 941.

6. 'Religion is partly fundamental and immutable, partly circumstantial and mutable.' 'A Short Schem of the True Religion', Keynes MS 7, in Cohen and Westfall, *Newton: Texts*, p. 344.

7. Quoted in Westfall, *Never at Rest*, p. 348.

8. Scholars agree that no ancient Greek texts include the phrase *these three are one*. Modern English translations have instead (typically) *the three are in agreement*.

9. 'Two Notable Corruptions of Scripture'; *Corres* III: 83; etc.

10. Quoted in Dobbs, *Foundations of Newton's Alchemy*, p. 164. Also Jan Golinski, 'The Secret Life of an Alchemist', in Fauvel et al., *Let Newton Be!*

11. By the end of his life, a few people knew, including William Whiston, his successor as Lucasian Professor at Cambridge. Whiston was stripped of the professorship and tried for heresy because he made his own Arianism public. He had received the post because of Newton's patronage; then Newton refused him membership in the Royal Society because – Whiston believed – 'they durst not choose a Heretick'. Whiston said of his patron, 'He was of the most fearful, cautious, and suspicious temper, that I ever knew.' *Memoirs*, pp. 250 f.

Westfall notes (*Never at Rest*, p. 318) that Isaac Barrow had gone so far as to write a 'Defense of the Blessed Trinity', and his successor as Master of Trinity College vowed to 'batter the atheists and then the Arians . . .'

By the time of his death, rumours of Newton's Arianism had circulated, but his friends and then his biographers heartily denied them. E.g. Stukeley (*Memoirs*, p. 71): 'Several people of heretical and unsettled notions, particularly those of Arian principles, have taken great pains to inlist Sir Isaac into their party, but that with as little justice as the anti-christians.'

12. Newton seems to have drafted this dispensation himself. No one knows how he gained royal approval; perhaps Barrow interceded for him.

13. Yahuda MS 14, quoted in Westfall, *Never at Rest*, p. 315.

14. Ibid., p. 317 n.

15. Westfall, 'Newton's Theological Manuscripts', in Bechler, *Contemporary Newtonian Research*, p. 132.

16. 'A Short Schem of the True Religion', Keynes MS 7, in Cohen and Westfall, *Newton: Texts*, p. 345.

II: FIRST PRINCIPLES

1. Add MS 404.

2. But this was not 'Halley's Comet'. That came next, in 1682. It was not till 1696 – having ingested the revelations of Newton's *Principia* and having obtained data from a now-hostile Flamsteed – that Halley calculated its path as an ellipse rather than a parabola and predicted its return every seventy-six years.

3. Andrew P. Williams, 'Shifting Signs: Increase Mather and the Comets of 1680 and 1682', *Early Modern Literary Studies* 1: 3 (December 1995).

4. Flamsteed to Crompton for Newton, 15 December 1680, *Corres* II: 242.

5. Schaffer, 'Newton's Comets and the Transformation of Astrology', p. 224. Indeed, Hooke had been suggesting that comets might orbit the sun with periods of many decades and that the paths of comets might be bent into a curve by the attractive power of the sun. Pepys, *Diaries*, 1 March 1665; Hooke, *Cometa*, 1678.

6. Flamsteed to Crompton, 3 January 1681, *Corres* II: 245.

7. Flamsteed to Crompton, 12 February 1681, *Corres* II: 249.

8. Flamsteed to Halley, 17 February 1681, *Corres* II: 250.

9. Newton to Crompton for Flamsteed, 28 February 1681, *Corres* II: 251. It is now clear that the data available to Newton were riddled with errors and inconsistencies, some even caused by confusion over calendar differences.

10. 'The only way to releive this difficulty in my judgmt is to suppose the Comet to have gone not between the ☉ and the Earth but to have fetched a compass about the ☉.' Ibid.

11. Hooke to Newton, 24 November 1679, *Corres* II: 235.

12. Newton to Hooke, 28 November 1679, *Corres* II: 236.

13. *An Attempt to Prove the Motion of the Earth by Observations* (London:

John Martyn, 1674). Hooke implied, but did not state mathematically, that gravity was inversely proportional to distance: 'these attractive powers are so much more powerful in operating, by how much nearer the body wrought upon is to their own Centers'.

14. Newton to Hooke, 28 November 1679, *Corres* II: 236. Hooke took this for a lie: 'He here pretends he knew not H's hypothesis,' he wrote on the letter. And he was right. Newton admitted it to Halley in 1686. Cf. Koyré, 'An Unpublished Letter of Robert Hooke to Isaac Newton', in *Newtonian Studies*, p. 238 n., and Westfall, *Never at Rest*, p. 383 n.

15. Newton to Hooke, 28 November 1679, *Corres* II: 236.

16. The discussion that followed shows, as Koyré says, 'the level of understanding – or lack of understanding – of even the best minds of the time'. Christopher Wren suggested shooting a bullet almost straight up, but 'round every way,' to see if the bullets all fall in a perfect circle. Flamsteed said it was well known that a ball shot directly upwards would not fall back into 'the mouth of the piece'; he suggested an angle of 87 degrees. Koyré, *Newtonian Studies*, p. 246.

17. Hooke to Newton, 9 December 1679. Newton had made a double error, in fact, because he also noted that such an object dropped in the Northern Hemisphere would tend southward as well as eastward. But there are complexities. Hooke was assuming a vacuum; as Newton later pointed out, the path through a resisting medium such as air would in fact be a spiral reaching the earth's centre. Also, neither man was ready (at first) to work out what it meant, gravitationally, to consider the earth's mass as spread through a sphere extending outside the path of the falling object, rather than concentrated at a central point. Koyré, *Newtonian Studies*, p. 248, and *Corres* II: 237.

18. He later told Halley, 'I refused his correspondence, told him I had laid Philosophy aside, ... expected to hear no further from

him, could scarce perswade my self to answer his second letter; did not answer his third.' Newton to Halley, 20 June 1686, *Corres* II: 288.

19. Hooke placed the centre of the earth incorrectly at the ellipse's centre, rather than at a focus. Hooke to Newton, 9 December 1679, *Corres* II: 237; Newton to Hooke, 13 December 1679, *Corres* II: 238.

A thorough and persuasive analysis of these diagrams and what they reveal about Newton's understanding of the possibilities – backward to his first mathematics on curvature and forward to the *Principia* – is J. Bruce Brackenridge and Michael Nauenberg, 'Curvature in Newton's Dynamics', in Cohen and Smith, *Cambridge Companion to Newton*.

20. Hooke to Newton, 6 January 1680, *Corres* II: 239.

21. Hooke to Newton, 17 January 1680, *Corres* II: 240.

22. 'Mr. Hook then sd that he had it, but that he would conceale it for some time that others triing and failing, might know how to value it, when he should make it publick.' Halley to Newton, 29 June 1686, *Corres* II: 289.

23. Add MS 3965, *De Motu Corporum*, in Hall and Hall, *Unpublished Scientific Papers*, p. 241.

24. *De Motu Corporum in Gyrum*, in Herivel, *Background to Newton's Principia*, pp. 257–89.

25. Flamsteed to Newton, 27 December 1684, *Corres* II: 273. Flamsteed did eventually see it.

26. Flamsteed to Newton, 27 December 1684, *Corres* II: 273, and Newton to Flamsteed, 12 January 1685, *Corres* II: 276.

27. Humphrey Newton's recollections, quoted in Westfall, *Never at Rest*, p. 406.

28. *Principia* 382.

29. '. . . the manner of expression will be out of the ordinary and purely mathematical . . . Accordingly those who there interpret these words as referring to the quantities being measured do violence to the Scriptures. And they no less corrupt mathematics and philosophy . . .' *Principia* 414.

30. *Principia* 408.

12: EVERY BODY PERSEVERES

1. Birch, *History of the Royal Society,* 4: 480.

2. Humphrey Newton (no relation).

3. Birch, *History of the Royal Society,* 4: 480.

4. Halley to Newton, 22 May 1686, *Corres* II: 285.

5. Newton to Halley, 27 May, 20 June, 14 July, and 27 July 1686, *Corres* II: 286, 288, 290, 291.

6. Westfall, *Never at Rest,* p. 449. Having obliterated Hooke, he gave early and prominent mention to 'Sir Christopher Wren, Dr. John Wallis, and Mr. Christiaan Huygens, easily the foremost geometers of the previous generation'. *Principia* 424.

7. Newton to Halley, 20 June 1686, *Corres* II: 288.

8. Francis Willoughby and John Ray, *Historia Piscium* (London: John Martyn, printer to the Royal Society, 1678).

9. Halley to Newton, 24 February 1687, *Corres* II: 302.

10. Halley to Newton, 5 July 1687, *Corres* II: 309.

11. *Phil. Trans.* 16: 291.

12. *Principia* 416–17.

13. Cf. J. R. Milton, 'Laws of Nature', in Garber and Ayers, *Cambridge History of Seventeenth-Century Philosophy,* p. 680. The practice of naming 'laws' after their scientific discoverers did not exist; it was born here. Kepler's laws antedate Newton's, but *Kepler's laws* is an eighteenth-century back-formation.

14. *Natura valde simplex est et sibi consona.* 'Conclusio' (Add MS 4005), in Hall and Hall, *Unpublished Scientific Papers,* p. 333.

15. Modern students of physics, with the calculus in their arsenal,

often find it simple to derive a result of Newton's by calculus yet difficult to understand the same result in the geometrical terms Newton employed in the *Principia*. Newton foresaw this himself. Thirty years later, he gave an anonymous account, writing of himself in the third person:

By the help of the new *Analysis* Mr. Newton found out most of the Propositions in his *Principia Philosophiæ*, but because the Ancients for making things certain admitted nothing into geometry before it was demonstrated synthetically, that the Systeme of the Heavens might be founded upon good Geometry. And this makes it now difficult for such unskillful men to see the Analysis by which those Propositions were found out. *Phil. Trans.* 29 (1715): 206.

Newton made this and similar self-serving claims about his use of the calculus in the course of his dispute with Leibniz about which of them had invented it. Scholars have debated it endlessly. They have found nothing like a discarded draft of the *Principia* in terms of the new analysis.

16. *Principia* 442.

17. *Principia* 590.

18. Recalled after Newton's death by Conduitt, at second or third hand. Keynes MS 130.6.

19. *Principia* 793 and Keynes MS 133.

20. *Principia* 790.

21. *Principia* 803.

22. Here and in several other calculations, he was not above manipulating the numbers to produce the appearance of exactitude. No one called his bluff. Galileo, in a comparable position, had elected to stay away from precise numerical calculations, saying that such vagaries as air resistance do not 'submit to fixed laws and exact description ... It is necessary to cut loose from such difficulties.' Newton, by contrast, set himself, and science, the obligation to exclude nothing and calculate everything. As Westfall says, 'So completely has modern physical science modeled itself on the *Principia* that we can scarcely realize how unprecedented such calculations were.' It was impossible, given the available data, and sometimes he cheated. Westfall, 'Newton and the Fudge Factor', *Science* 179 (23 February 1973): 751. Also Nicholas Kollerstrom, 'Newton's Lunar Mass Error', *Journal of the British Astronomical Association* 95 (1995): 151. For another example of what Whiteside calls 'the delicate art of numerical cookery', see *Math* VI: 508–36.

23. *Principia* 807.

24. *Principia* 806.

25. *Principia* 814.

26. *Principia* 829.

27. Add MS 3965, 'De motu corporum', in Hall and Hall, *Unpublished Scientific Papers*, p. 281.

28. *Principia* 875–78 and 839. There was nothing conclusive in this data, but Newton did not pass it by. He did not restrict himself to idealised tides but tried to consider the geography of estuaries and rivers. He studied the map of Batsha Harbour, with multiple inlets and open channels, reaching the China Sea and the Indian Ocean, and worked out a theory of wave interference that could account for the data. I. Bernard Cohen, 'Prop. 24: Theory of the Tides; The First Enunciation of the Principle of Interference", in *Principia* 240; Ronan, *Edmond Halley*, pp. 69f.

29. Galileo, *Dialogue*, pp. 445 and 462.

30. These explicitly became rules in the second edition; in the first, they were called 'hypotheses'. *Principia* 794–96. There were four rules in all; the others were:

Those qualities of bodies that cannot be increased or diminished and that belong to all bodies on which experiments can be made should be taken as qualities of all bodies universally.

In experimental philosophy, propositions gathered from phenomena by induction should be considered either exactly or very nearly true notwithstanding any contrary hypotheses, until yet other phenomena make such propositions either more exact or liable to exceptions.

31. Quoted in Westfall, *Never at Rest*, p. 464.
32. 'I do not feign hypotheses' is the most popular solution to one of history's most debated translation problems: '*Hypotheses non fingo*'. A reasonable alternative is 'frame'. Either way, Newton always gets credit for this phrase, but he did not invent it. Henry Oldenburg (for example) had described the Royal Society's virtuosi as those 'who, neither feigning nor formulating hypotheses of nature's actions, seek out the thing itself'. Oldenburg to Francisco Travagino, 15 May 1667.
33. *Principia* 943.
34. *Principia* 382.

13: IS HE LIKE OTHER MEN?

1. 'Aphorisms Concerning the Interpretation of Nature and the Kingdom of Man', Bacon, *Novum Organum*, p. 43.
2. 'This incomparable Author having at length been prevailed upon to appear in publick, has in this Treatise given a most notable instance of the extent of the powers of the Mind . . .' *Phil. Trans.* 186: 291.
3. Halley to King James II, July 1687, *Corres* II: 310. Whatever James did with his copy, it did not survive.
4. Halley, 'The true Theory of the Tides, extracted from that admired Treatise of Mr. Isaac Newton, Intituled, Philosophiæ Naturalis Principia Mathematica', *Phil. Trans.* 226: 445, 447.
5. Untitled draft, *Corres* II: 301.
6. Newton to John Covel, 21 February 1689, *Corres* III: 328.

7. Godfrey Kneller, 1689. See frontispiece.

8. Newton to a Friend, 14 November 1690, *Corres* III: 358. 'Yes truly those Arians were crafty Knaves that could conspire so cunningly & slyly all over the world at once.'

9. Pepys to Newton, 22 November 1693, *Corres* III: 431. Pepys was more interested than most in arithmetical matters; he had learned multiplication at the age of twenty-nine with the help of a ship's mate. Thomas, 'Numeracy in Early Modern England', pp. 111-12.

10. Newton to Locke (draft), December 1691, *Corres* III: 377.

11. Defoe, *A Journal of the Plague Year*, p. 1.

12. Johns, *The Nature of the Book*, pp. 536–37.

13. *Bibliothèque Universelle et Historique* (March 1688, probably by Locke himself), *Acta Eruditorum* (June 1688), and *Journal des Sçavans* (August 1688).

14. Keynes MS 130.5, quoted in Westfall, *Never at Rest*, p. 473.

15. Newton to Bentley, 25 February 1693, *Corres* III: 406.

16. Draft of the General Scholium (section IV, no. 8, MS C), in Hall and Hall, *Unpublished Scientific Papers*, p. 90.

17. Newton to Bentley, 10 December 1692, *Corres* III: 398.

18. *Corres* III: 395.

19. 'The Rise of the Apostasy in Point of Religion', Yehuda MS 18, Jewish National and University Library, Jerusalem.

20. The particulars of Newton's breakdown will for ever inspire debate and speculation. As for the fire, most believe Newton lost some papers to fire in the late seventies; Westfall goes further and suggests, 'There may . . . have been a fire – another fire, as it appears to me – which could well have distracted him when he was already in a state of acute tension. Charred papers survive from the 1690s, though it is difficult to fit them . . .' *Never at Rest*, p. 538. A popular legend involving a dog called Diamond and a candle (cf. *Bartlett's Familiar Quotations*) is surely apocryphal. As for the involuntary restraint: no. As for mercury poisoning: he did suffer symptoms such as insomnia and apparent paranoia, but he lacked others, and these were temporary; modern tests of his hair did reveal toxic mercury levels, but the hair cannot be dated. Some of the debate plays out in Spargo and Pounds, 'Newton's "Derangement of the Intellect"'; Johnson and Wolbarsht, 'Mercury Poisoning: A Probable Cause of Isaac Newton's Physical and Mental Ills'; Ditchburn, 'Newton's Illness of 1692–3'; and Klawans, *Newton's Madness*.

Whiteside has summarised the scholarly state of affairs: 'Where scholars have, from the pedestals of their own stand-points, bickered ceaselessly this past century and a half over the possible causes and long-term after-effects of Newton's undeniable breakdown ... we would be foolish to attempt any definitive assessments when the extant record offers but a blurred glimpse...' *Math* VII: xviii.

21. Newton to Pepys, 13 September 1693, and Newton to Locke, 16 September 1693; *Corres* III: 420 and 421.

22. Pepys to Millington, 26 September 1693, *Corres* III: 422.

23. Quoted by Whiteside, *Math* VII: 198.

24. David Gregory was the new professor of astronomy and an original proselytiser for the *Principia*. 'David Gregory's Inaugural Lecture at Oxford', *Notes and Records of the Royal Society* 25 (1970): 143–78.

25. Whiston, *Memoirs*, p. 32.

26. *Oeuvres de Huygens* XXI: 437, quoted in Westfall, *Force in Newton's Physics*, p. 184, and cf. Guerlac, *Newton on the Continent*, p. 49.

27. Guerlac, *Newton on the Continent*, p. 52.

28. Unpublished draft, quoted in Hall, *Philosophers at War*, p. 153.

29. Leibniz to Newton, 7 March 1693, *Corres* III: 407. It was their first contact since the brief correspondence sixteen years before.

30. Memoranda by David Gregory, *Corres* IV: 468, and Flamsteed's recollection, *Corres* IV: 8 n.; Newton to Flamsteed, 7 January 1694, *Corres* IV: 473.

31. Newton to Flamsteed, 20 July 1695, *Corres* IV: 524.

32. Newton to Flamsteed, 6 January 1699, *Corres* IV: 601.

33. To Newton Flamsteed wrote: 'I have somtimes told some ingenious men that more time and observations are required to perfect the Theory but I found it was represented as a little peice of detraction *which I hate* . . . I wonder that *hints* shoud drop from your pen, as if you Lookt on my business as *trifling.*' 10 January 1699, *Corres* IV: 604.

34. Nicholas Kollerstrom's computer-assisted analysis, *Newton's Forgotten Lunar Theory*, is definitive. Kollerstrom judges the method, as employed by Halley, as accurate enough to have won a £10,000 prize established by Parliament in 1714.

35. Westfall, *Never at Rest*, p. 550. He did retain his professorship and salary, but he seldom visited Cambridge again, and 'as far as we know, he wrote not a single letter back to any acquaintance made during his stay'.

14: NO MAN IS A WITNESS IN HIS OWN CAUSE

1. Westfall, *Never at Rest*, p. 699.

2. The problem was to find the curve (the *brachistochrone*) along which a body descending to a given point under its own gravity will take the shortest time. (Roughly: the shape of the fastest track for a roller coaster.) Galileo had thought the curve of fastest descent would be a simple arc of a circle, which is certainly faster than a straight-line ramp. In fact it is the curve known as the cycloid.

Bernoulli had posed the problem with Newton in mind, as a challenge, in the context of the simmering calculus priority dispute. He addressed it to 'the very mathematicians who pride themselves that . . . they have not only penetrated most intimately the hiding-places of a more secret Geometry, but have even extended its limits in a remarkable way by their golden theorems' (quoted by Mandelbrote,

Footprints of the Lion, p. 76). Newton solved it the night it arrived, and to Whiteside ('Newton the Mathematician', in Bechler, *Contemporary Newtonian Research*, p. 122) this feat was evidence of the deterioration of his mathematical powers in old age: 'A couple of years earlier his method of "maxima & minima in infinitesimals" would have detected that this is the cycloid in a few minutes, not the twelve hours he in his rustiness then took.'

3. Westfall, *Never at Rest*, p. 721.

4. Valentin Boss, *Newton and Russia*.

5. Hoppit, *A Land of Liberty?*, p. 186.

6. '. . . thou hast ordered all things in measure and number and weight'. Wisdom of Solomon 11: 20.

7. Petty, *Political Arithmetick*.

8. Newton upon becoming Warden was obliged to swear an oath: 'You will not reveal or discover to any person or persons whatsoever the new Invention of Rounding the money & making the edges of them with letters or grainings or either of them directly or indirectly. So help you God.' *Corres* IV: 548.

9. '. . . only 400*lib* per annum with a house of 40*lib* per annum & his perquisites are only 3*lib* 12*s* per annum . . . so small . . . not to support the authority of the office.' *Corres* IV: 551.

10. On the matter of Newton and crimson, no one has been more eloquent than Richard de Villamil in 1931 (*Newton the Man*, pp. 14–15), after analysing his household inventory:

. . . crimson mohairs nearly everywhere. Newton's own bed was a 'crimson mohair bed,' with 'crimson Harrateen' bed-curtains' . . . 'crimson mohair hangings' . . . a 'crimson sattec.' In fact, there is no other colour referred to in the 'Inventary' but crimson. This living in what I may call an 'atmosphere of crimson' is probably one of the reasons why Newton became rather irritable towards the end of his life.

11. Newtonians struggled with euphemisms for this relationship even into the twentieth century ('about the exact nature of [their] friendship there has been unseemly speculation,' wrote Andrade in 1947). When Halifax died, in 1715, he left Barton a bequest of more than £20,000 – 'for her excellent conversation,' Flamsteed wrote maliciously. There was gossip (though this mangled the sequence of events) that the connection had facilitated Newton's appointment to the Mint; Voltaire spread it most famously: 'The infinitesimal calculus and gravitation would have been of no use without a pretty niece' (*Lettres Philosophiques*, letter 21).

Then again, Newton's Freudian biographer Frank Manuel avoided euphemism altogether, choosing to view Catherine as an incarnation of Hannah: 'In the act of fornication between his friend Halifax and his niece was Newton vicariously having carnal intercourse with his mother?' Manuel, *Portrait*, p. 262.

12. Montague to Newton, 19 March 1696, *Corres* IV: 545.

13. China, for example, placed a higher value on silver than Europe did, and arbitrage ensued. 'Our silver must go to China till gold is dearer there or cheaper with us,' Newton wrote. 'The trade for their gold must greatly increase our coin, being a profit to the nation . . .' Craig, *Newton at the Mint*, p. 43.

14. 'Observations concerning the Mint', *Corres* IV: 579.

15. Newton and Ellis to Henry St John, September 1710, *Corres* V: 806.

16. Signed, 'Your near murderd humble Servant W. Chaloner'. Chaloner to Newton, 6 March 1699, *Corres* IV: 608.

17. Memorandum, 'Of the assaying of Gold and Silver, the making of indented Triall-pieces, and trying the moneys in the Pix', Mint Papers (Public Record Office, Kew), I: f. 109. 'A Scheme of a Commission for prosecuting Counterfeiters & Diminishers of the current coyn', manuscript, Pierpont Morgan Library.

18. He issued this bill first in April and then in December.

19. Wallis to Newton, *Corres* IV: 503 and 567. Wallis added, 'I should say the same about many things you keep hidden, of which I am not yet aware.'

20. Stukeley, *Memoirs*, p. 79.

21. A Latin version of the *Opticks* appeared two years later, in 1706 – long before the first English version of the *Principia*, in 1729.

22. *Advertisement* to *Opticks*, first edition.

23. *Opticks*, Query 29, p. 370.

24. These are still called Newton's rings. Nevertheless, reluctant though Newton was to admit it, the origins of this investigation lay in Hooke's *Micrographia*.

25. *Opticks*, book II, part 3, proposition XIII, p. 280. Cf. Westfall, 'Uneasily Fitful Reflections on Fits of Easy Transmission', in Palter, *Annus Mirabilis*, pp. 88–104.

26. E.g. *Opticks*, p. 376. Newton's grandest metaphysical speculation – particularly the credo of Query 31 – did not appear full-blown in the first edition, but evolved, beginning with the Latin edition of 1706.

27. *Opticks*, p. 394.

28. *Opticks*, p. 404.

29. Francis Hauksbee, a former assistant of Robert Boyle, and then John Theophilus Desaguliers, later a renowned populariser of Newton in prose and verse.

30. Quoted in Heilbron, *Physics at the Royal Society*, p. 65.

31. *Opticks*, p. 405.

32. *Opticks*, pp. 399–400.

33. The first French translation did not appear until 1720. Even so this preceded by thirty-six years the first – and only – French translation of the *Principia*, by Gabrielle-Émilie Le Tonnelier de Breteuil, Marquise du Châtelet, Voltaire's friend and lover ('she was a great man whose only fault was in being a woman'). Her name, and not Newton's, appeared on the title page. It was Châtelet who said of Cartesianism: 'It is a house collapsing into ruins, propped up on every side . . . I think it would be prudent to leave.'

34. Guerlac, *Newton on the Continent*, p. 51 n.

35. 'I have even noticed certain things from which it appears that Dynamics, or the law of forces, are not deeply explored by him.' Leibniz to J. Bernoulli, 29 March 1715, *Corres* IV: 1138. Newton was well, if belatedly, aware of the danger of *sensorium*, and he backtracked in revising this passage.

36. Alexander, *Leibniz-Clarke Correspondence*, p. 30. Howard Stein suggests that if Leibniz had understood the 'incoherence' of his relativism, he would have been better equipped to appreciate gravitation. 'Newton's Metaphysics', in Cohen and Smith, *Cambridge Companion to Newton*, p. 300.

37. The so-called Epistola Posterior, Newton to Oldenburg, 24 October 1676, *Corres* II: 188. Cf. *Principia* 651 n. The key is in Add MS 4004.

38. '. . . which without our differential calculus no one could attack with such ease.' *Acta Eruditorum*, May 1684, trans. D. J. Struik, in Fauvel and Gray, *History of Mathematics*, p. 434.

39. Newton's letters to Leibniz first appeared in John Wallis's third volume of *Opera Mathematica* in 1699 – a deliberate marshalling of evidence. Barrow had sent Collins Newton's *De Analysi per Æquationes Infinitas* in 1669, and Collins, before returning it, had made at least one copy – which he showed Leibniz in 1676.

40. John Keill, *Phil. Trans.* 26 (1709), quoted by Westfall, *Never at Rest*, p. 715.

41. *Corres* V: xxiv.

42. 'An Account of the Book Entituled *Commercium Epistolicum, Collinii et Aliorum, de Analysi Promota*', *Phil. Trans.* 342 (February 1715): 221.

43. Ibid., pp. 205 and 206.

44. Ibid., pp. 216, 209 and 208.

45. Ibid., pp. 223–24.

46. As L. J. Russell put it: 'You might at any moment hit on a simple substitution, e.g. in an algebraic equation, or in a summation of a series, that would lead to a new general method . . . Sometimes even the hint that someone had discovered a method for solving a particular problem was enough to set you looking in the right direction for solving it, and you could solve it too. In such a situation, what is needed is a general clearing house of publicity.' 'Plagiarism in the Seventeenth Century, and Leibniz', in Greenstreet, *Isaac Newton*, p. 135.

47. Leibniz's symbols did not map neatly onto the notation Newton had devised for his own use, dotted letters for fluxions and various alternatives for fluents, and the consequence was that British and Continental mathematics diverged throughout the eighteenth century. Finally, in the nineteenth, Leibniz's differential notation prevailed over the dots even in England.

48. Lenore Feigenbaum, 'The Fragmentation of the European Mathematical Community', in Harman and Shapiro, *Investigation of Difficult Things*, p. 384. She also quotes Whiteside, calling the controversy 'a long-festering boil [that] polluted the whole European world for a decade afterwards with the corruption of its discharging pus'. *Math* VIII: 469.

49. Baily, *Account of the Revd John Flamsteed*, p. 294. Flamsteed died soon after, having been Royal Astronomer forty-five years, and Halley took his place.

50. *Math* VII: xxix.

51. Leibniz to Rémond de Montmort, 19 October 1716, quoted in Manuel, *Portrait*, p. 333.

15: THE MARBLE INDEX OF A MIND

1. Nicolson, *Science and Imagination*, p. 115.
2. Swift, *Gulliver's Travels*, III: 8.
3. *Letters on England*, No. 13, p. 67.

4. *Letters on England*, pp. 86 and 79–80. For a generation to come, Anglo-French rivalry coloured Newton's reception in France. He had been elected a foreign member of the Académie Royale in 1699 but had never acknowledged the honour or communicated with the academy. When French scientists meant 'Newtonians', they generally said, 'les anglais'.

5. Bernard le Bovier de Fontenelle, *The Elogium of Sir Isaac Newton* (London: Tonson, 1728), read to the Académie Royale des Sciences in November 1727; reprinted in Cohen, ed., *Papers and Letters*, pp. 444–74; based in turn on John Conduitt's 'Memoir', in *Isaac Newton: Eighteenth-Century Perspectives*, pp. 26–34. ('He had such a meekness and sweetness of temper . . . His whole life was one continued series of labour, patience, charity, generosity, temperance, piety, goodness, and all other virtues, without a mixture of any vice whatsoever.') In the contemporary manner, Fontenelle also embellished Newton's ancestry: 'descended from the elder branch of the family of Sir John Newton Baronet . . . Manor of Woolstrope had been in his Family near 200 years . . . Sir Isaac's Mother . . . was likewise of an ancient family . . .' In fairness to Fontenelle, he relied on a pedigree Newton had embellished himself, after being knighted.

The supposedly singular laugh is originally due to Humphrey Newton; Stukeley (*Memoirs*, p. 57) considered this at length and said that he had often seen Newton laugh and that 'he was easily made to smile, if not to laugh'.

6. Quoted in Paul Elliott, 'The Birth of Public Science'. p. 77.

7. '. . . Nature her self to him resigns the Field,/From him her Secrets are no more conceal'd.' *Gentleman's Magazine* I (February 1731): 64.

8. *Gentleman's Magazine* I (April 1731): 157.

9. *Epitaphs* (1730). Here Pope was serving a long, slow pitch to the twentieth-century wag who replied: 'It did not last: the Devil howling "Ho,/Let Einstein be," restored the status quo.' Koyré, 'The Newtonian Synthesis', in *Newtonian Studies*.

10. One of the observers, William Whiston, said he made enough money from eclipse lectures and 'the sale of my schemes before and after' to support his family for a year, and added: 'There happened to be a Mahometan envoy here from Tripoly, who at first thought we were distracted, by pretending to know so very punctually when God Almighty would totally eclipse the Sun; which his own mussel-men were not able to do ... When the eclipse came exactly as we foretold, he was asked again, what he thought of the matter now? His answer was, that he supposed we knew this by art magick.' *Memoirs*, p. 205.

11. George Gordon (London: W.W., 1719).

12. Not till 1890, if we believe the *OED*, and its first appearance was pejorative: '1890 *Athenæum* 19 July 92/2 [Mercier] declared Newtonianism to be the "most absurd scientific extravagance that has ever issued from the human imagination".'

13. *Sir Isaac Newton's Philosophy Explain'd for the Use of the Ladies* (London: E. Cave, 1739), p. 231.

14. Socolow, 'Of Newton and the Apple', *Laughing at Gravity*, p. 7.

15. Haydon's *Autobiography* (1853), quoted in Nicolson, *Newton Demands the Muse*, p. 1; and Penelope Hughes-Hallett, *The Immortal Dinner: A Famous Evening of Genius and Laughter in Literary London, 1817* (London: Viking, 2000).

16. Keats, *Lamia* (1819).

17. Shelley, *Queen Mab*, V: 143–45. He read Newton carefully and with understanding. 'We see a variety of bodies possessing a variety of powers: we merely know their effects; we are in a state of ignorance with respect to their essences and causes. These Newton calls the phenomena of things; but the pride of philosophy is unwilling to admit its ignorance of their causes.' *Notes to Queen Mab*, VII.

18. Wordsworth, *The Prelude*, III.

19. Blake, *The Book of Urizen*, I.

20. Blake, 'Annotations to the works of Sir Joshua Reynolds'.

21. Blake, 'On the Virginity of the Virgin Mary & Johanna Southcott' (*Satiric Verses & Epigrams*). Also: 'To teach doubt & Experiment / Certainly was not what Christ meant.' *The Everlasting Gospel.*

22. Blake, *Jerusalem*, Chapter I.

23. Brewster, *Life of Sir Isaac Newton*, p. 271.

24. Byron, *Don Juan*, Canto X.

25. Burtt, *Metaphysical Foundations*, pp. 203, 303.

26. *Principia* (Motte), p. 192.

27. As Clifford Truesdell puts it ('Reactions of Late Baroque Mechanics to Success, Conjecture, Error, and Failure in Newton's *Principia*', in Palter, *Annus Mirabilis*, p. 192): 'Newton relinquished the diplomatic immunity granted to nonmathematical philosophers, chemists, psychologists, etc., and entered into the area where an error is an error even if it is Newton's error; in fact, all the more so because it is Newton's error.'

28. Cohen, *Revolution in Science*, pp. 174–75.

29. Steven Weinberg, 'The Non-Revolution of Thomas Kuhn', in *Facing Up*, p. 197: 'Kuhn knew very well that physicists today go on using the Newtonian theory of gravitation and motion ... We certainly don't regard Newtonian and Maxwellian theories as simply false, in the way that Aristotle's theory of motion or the theory that fire is an element are false.'

30. Quoted in Fara, *Newton*, p. 256.

31. Kuhn, *Structure of Scientific Revolutions*, p. 108.

32. Einstein's space-time was not, therefore, that of Leibniz and other contemporary anti-Newtonians. As H. G. Alexander notes, Leibniz's critique of absolute space and time in no way anticipated Einstein's: 'Leibniz's fundamental postulate is that space and time are unreal. No one therefore would have rejected more strongly than he a theory which ascribes properties to space-time.' *The Leibniz-Clarke Correspondence*, Introduction, p. lv. Or, as Howard Stein puts it: 'In spite of the fact that absolute space and absolute time have been abandoned, and the geometric structure of space-time has proved to be *interdependent with* the distribution of matter ... it remains necessary to regard space-time and its geometry as having a status as "real" as that of matter . . . On this *general* score – although certainly not in *detail* – Newton was, in the eyes of our own science, "right" to take space and time as fundamental entities.' 'Newton's Metaphysics', in Cohen and Smith, *Cambridge Companion to Newton*, p. 292.

33. Einstein, 'What Is the Theory of Relativity?' *Times* of London, 28 November 1919, reprinted in *Out of My Later Years*, p. 58a. And as he put it a few years later (1927): 'We have to realize that before Newton there existed no self-contained system of physical causality which was somehow capable of representing any of the deeper features of the empirical world.' Einstein, 'The Mechanics of Newton and Their Influence on the Development of Theoretical Physics', in *Ideas and Opinions*, p. 277.

34. *Principia* (Motte), p. 8.

35. *Opticks* 374.

36. *Principia* 407.

37. *Opticks* 388–89.

38. Even here, in establishing this fundamental dictum of science, he allowed for the alternative possibility. His heirs and followers forgot, but he wrote: 'it may also be allow'd that God is able to create Particles of Matter of several Sizes and Figures, and in several Proportions to Space, and perhaps of different Densities and Forces, and thereby to vary the Laws of Nature, and make Worlds of several sorts in several Parts of the Universe. At least, I see nothing of Contradiction in all this.' *Opticks* 403–4.

39. 'Newton and the Twentieth Century – A Personal View', in Fauvel et al., *Let Newton Be!*, p. 244.

40. Scott Mandelbrote says: 'The causes of this are hard to fathom but may relate to the international situation, the sense that Cambridge already possessed all that mattered of Newton's papers, fatigue in a market that was already awash with books from Newton's library, or even disquiet at Lord Lymington's right-wing political views.' *Footprints of the Lion*, p. 137.

The total sale amounted to barely £9,000, including two portraits and the death mask. Most of the interest and the pre-sale publicity came from the United States. P. E. Spargo, 'Sotheby's, Keynes, and Yahuda: The 1936 Sale of Newton's Manuscripts', in Harman and Shapiro, *Investigation of Difficult Things*, pp. 115–34.

41. John Maynard Keynes, 'Newton the Man', in Royal Society, *Newton Tercentenary Celebrations*, p. 27. Freeman Dyson, who was there, describes Keynes's talk in *Disturbing the Universe* (New York: Harper & Row, 1979), pp. 8–9.

42. Edited by Conduitt, this became (after the *Principia* and *Opticks*) one of his first works published, the year after his death. To modern eyes it is, as Westfall declared plaintively, 'a work of colossal tedium ... read today only by the tiniest remnant who for their sins must pass through its purgatory'. Westfall, *Never at Rest*, p. 815.

43. Keynes MS 130.11; Brewster, *Life of Sir Isaac Newton*, p. 324.

44. 'Principles of Philosophy', manuscript fragment (c. 1703), Add MS 3970.3

45. Stukeley, *Memoirs*, pp. 25–26.

46. Inventory, 'Dom Isaaci Newton, Mil.', dated 5 May 1727, in de Villamil, *Newton the Man*, pp. 49–61.

Acknowledgements and Sources

I have meant to ground this book as wholly as possible in its time; in the texts. The diaspora of Newton's manuscripts began at his death, continued for more than three centuries, and has only lately been reversed. They are still widely scattered, but the Cambridge University Libraries have gathered much of the essential core holding, including much of Newton's own library, annotated by him. I am indebted to Adam J. Perkins and others for great courtesy. Documents are cited according to the Cambridge numbering scheme as Add MS (Additional Manuscripts) or Keynes MS (Keynes Collection at Kings College). I am grateful to Joanna Corden, Rafael Weiser, Silvie Merian, and their colleagues in the archives of the Royal Society of London, the Jewish National and University Library in Jerusalem (Yehuda MS), and the Pierpont Morgan Library in New York, and to the National Trust custodians at Woolsthorpe Manor, for access and knowledge.

For guidance, criticism, and correction, I owe special thanks to James Atlas, Cynthia Crossen, Peter Galison, Scott Mandelbrote, Esther Schor, Craig Townsend and Jonathan Weiner, as well as my agent, Michael Carlisle. Above all to my editor, Dan Frank.

PUBLISHED WORKS OF NEWTON

There is no such thing as *The Collected Works of Isaac Newton*. The Newton Project, at Imperial College, London, has long-term plans for the theological, alchemical and Mint writings. Meanwhile two monuments of scholarship are the collected correspondence and the collected mathematical papers:

Turnbull, Herbert W.; Scott, John F.; Hall, A. Rupert; and Tilling, Laura, eds. *The Correspondence of Isaac Newton* (cited as *Corres*). Seven volumes. Cambridge: Cambridge University Press, 1959–77.

Whiteside, D. T., ed. *The Mathematical Papers of Isaac Newton* (cited as *Math*). Eight volumes. Cambridge: Cambridge University Press, 1967–80.

Optical papers are in progress:

Shapiro, Alan E., ed. *The Optical Papers of Isaac Newton: The Optical Lectures 1670–1672*. Cambridge: Cambridge University Press, 1984.

I have depended on other essential texts collected or reproduced in these volumes:

Principia:

The Principia: Mathematical Principles of Natural Philosophy (cited as *Principia*). Translated by I. Bernard Cohen and Anne Whitman with the assistance of Julia Budenz. Berkeley: University of California Press, 1999.

Sir Isaac Newton's Mathematical Principles of Natural Philosophy and His System of the World. Translated by Andrew Motte (1729), revised by Florian Cajori. Berkeley: University of California Press, 1947.

Acknowledgements and Sources

Newton's Principia: The Central Argument: Translation, Notes, and Expanded Proofs. Dana Densmore and William H. Donahue. Santa Fe: Green Lion Press, 1995.

Opticks. Foreword by Albert Einstein. New York: Dover, 1952.

The Background to Newton's Principia: A Study of Newton's Dynamical Researches in the Years 1664–1684. John Herivel. Oxford: Oxford University Press, 1965.

Certain Philosophical Questions: Newton's Trinity Notebook. J. E. McGuire and Martin Tamny. Cambridge: Cambridge University Press, 1983.

Isaac Newton's Papers & Letters on Natural Philosophy. Edited by I. Bernard Cohen. Cambridge, Mass.: Harvard University Press, 1958.

The Janus Faces of Genius: The Role of Alchemy in Newton's Thought. Betty Jo Teeter Dobbs. Cambridge: Cambridge University Press, 1991.

Newton: Texts, Backgrounds, Commentaries. Edited by I. Bernard Cohen and Richard S. Westfall. New York: Norton, 1995.

The Preliminary Manuscripts for Isaac Newton's 1687 Principia, 1684–85. Introduction by D. T. Whiteside. Cambridge: Cambridge University Press, 1989.

The Unpublished First Version of Isaac Newton's Cambridge Lectures on Optics, 1670–1672. Introduction by D. T. Whiteside. Cambridge: University Library, 1973.

Unpublished Scientific Papers of Isaac Newton. Edited by A. Rupert Hall and Marie Boas Hall. Cambridge: Cambridge University Press, 1962.

OTHER PRIMARY AND SECONDARY SOURCES

The authoritative scientific biography is Richard S. Westfall's *Never at Rest* (Cambridge: Cambridge University Press, 1980). He offered a salutary warning to all who follow: 'The more I have studied him, the more Newton has receded from me . . . Only another Newton could hope fully to enter into his being, and the economy of the human enterprise is such that a second Newton would not devote himself to the biography of the first.'

Adair, John. *By the Sword Divided: Eyewitness Accounts of the English Civil War*. Bridgend, UK: Sutton, 1998.

Alexander, Henry Gavin, ed. *The Leibniz-Clarke Correspondence*. Manchester: Manchester University Press, 1956.

Algarotti, Francesco. *Sir Isaac Newton's Philosophy Explain'd For the Use of the Ladies. In Six Dialogues on Light and Colours*. London: E. Cave, 1739.

Andrade, Edward Neville da Costa. 'Newton's Early Notebook'. *Nature* 135 (1935): 360.

——. *Sir Isaac Newton: His Life and Work*. New York: Macmillan, 1954.

Arbuthnot, John. *An Essay on the Usefulness of Mathematical Learning in a Letter from a Gentleman in the City to His Friend in Oxford*. Oxford: The Theater, 1701.

Aubrey, John. *Brief Lives*. Edited by Oliver Lawson Dick. London: Secker & Warburg, 1949.

Ault, Donald D. *Visionary Physics: Blake's Response to Newton*. Chicago: University of Chicago Press, 1974.

Bacon, Francis. *The Essays, or Councils, Civil and Moral*. London: H. Clark, 1706.

——. *Novum Organum*. Translated and edited by Peter Urbach and John Gibson. Chicago: Open Court, 1994.

———. *The Works of Francis Bacon: Baron of Verulam, Viscount St. Alban, and Lord High Chancellor of England*. Edited by James Spedding, Robert L. Ellis and Douglas D. Heath. New York: Garrett Press, 1968.

Baily, Francis. *An Account of the Revd John Flamsteed, the First Astronomer-Royal, Compiled from His Own Manuscripts and Other Authentic Documents*. Reprint of the 1835 edition. London: Dawsons, 1966.

Banville, John. *The Newton Letter: A Novel*. Boston: David R. Godine, 1972.

Bate, John. *The Mysteryes of Nature and Art*. Third edition. London: Andrew Crooke, 1654.

Bechler, Zev, ed. *Contemporary Newtonian Research*. Dordrecht: D. Reidel, 1982.

Ben-Chaim, Michael. 'Newton's Gift of Preaching,' *History of Science* 36: 269-98 (September 1998).

Birch, Thomas. *The History of the Royal Society of London*. Four volumes. Facsimile of the London edition of 1756–57. Introduction by A. Rupert Hall. New York: Johnson, 1968.

Blake, William. *The Complete Poetry and Prose of William Blake*. Edited by David F. Erdman. Berkeley: University of California Press, 1982.

Blay, Michel. *Reasoning with the Infinite: From the Closed World to the Mathematical Universe*. Translated by M. B. DeBevoise. Chicago: University of Chicago Press, 2000.

Boss, Valentin. *Newton and Russia: The Early Influence, 1698–1796*. Cambridge, Mass.: Harvard University Press, 1972.

Boyle, Robert. *Experiments and Considerations Touching Colours*. London: Henry Herringman, 1664.

———. *The Sceptical Chymist: or Chymico-physical Doubts & Paradoxes*. London: Cadwell, 1661.

Brewster, David. *The Life of Sir Isaac Newton: The Great Philosopher*, revised and edited by W. T. Lynn. London: Gall & Inglis, 1855.

Broad, C. D. *Sir Isaac Newton*. London: Oxford University Press, 1927.

Bronowski, Jacob. *Magic, Science, and Civilization*. New York: Columbia University Press, 1978.

Buckley, Michael J. *Motion and Motion's God*. Princeton: Princeton University Press, 1971.

Burke, John G., ed. *The Uses of Science in the Age of Newton*. Berkeley: University of California Press, 1983.

Burton, Robert. *The Anatomy of Melancholy*. Edited by Floyd Dell and Paul Jordan-Smith. New York: Tudor, 1927.

Burtt, Edwin Arthur. *The Metaphysical Foundations of Modern Physical Science*. Second edition. London: Routledge & Kegan Paul, 1932.

Capp, Bernard. *Astrology and the Popular Press: English Almanacs 1500–1800*. London: Faber & Faber, 1979.

Castillejo, David. *The Expanding Force in Newton's Cosmos as Shown in His Unpublished Papers*. Madrid: Ediciones de Arte y Bibliofilia.

Challis, C. E., ed. *A New History of the Royal Mint*. Cambridge: Cambridge University Press, 1992.

Chapman, Allan. 'England's Leonardo: Robert Hooke and the art of experiment in Restoration England', *Proceedings of the Royal Institution of Great Britain* 67 (1996): 239–75.

Christianson, Gale E. *In the Presence of the Creator: Isaac Newton and His Times*. New York: Free Press, 1984.

——. *Isaac Newton and the Scientific Revolution*. Oxford: Oxford University Press, 1996.

Clark, David H.; and Clark, Stephen P. H. *Newton's Tyranny: The Suppressed Scientific Discoveries of Stephen Gray and John Flamsteed*. New York: W. H. Freeman, 2000.

Clay, C. G. A. *Economic Expansion and Social Change: England 1500–1700*. Two volumes. Cambridge: Cambridge University Press, 1984.

Cohen, H. Floris. 'The Scientific Revolution: Has There Been a British View? A Personal Assessment'. *History of Science* 37 (March 1999).

Acknowledgements and Sources

Cohen, I. Bernard. *The Birth of a New Physics.* New York: Norton, 1983.

———. *Franklin and Newton.* Philadelphia: American Philosophical Society, 1956.

———. 'Newton in the Light of Recent Scholarship'. *Isis* 51 (December 1960): 489–514.

———. 'Newton's Second Law and the Concept of Force in the *Principia*'. In Palter, *Annus Mirabilis*, 143–85.

———. 'Notes on Newton in the Art and Architecture of the Enlightenment'. *Vistas in Astronomy* 22 (1979): 523–37.

———. *Revolution in Science.* Cambridge, Mass: Harvard University Press, 1985.

Cohen, I. Bernard; and Smith, George E., eds. *The Cambridge Companion to Newton.* Cambridge: Cambridge University Press, 2002.

Collier, Arth. *Clavis Universalis, or, a New Inquiry after Truth, being a demonstration of the Non-Existence, or Impossibility, of an External World.* London: Robert Gosling, 1713.

Collingwood, R. G. *The Idea of Nature.* Oxford: Oxford University Press, 1945.

Cook, Alan. *Edmond Halley: Charting the Heavens and the Seas.* Oxford: Oxford University Press, 1998.

Costello, William T. *The Scholastic Curriculum at Early Seventeenth-Century Cambridge.* Cambridge, Mass.: Harvard University Press, 1958.

Couth, Bill, ed. *Grantham during the Interregnum: The Hall Book of Grantham, 1641–1649.* Woodbridge, UK: Lincoln Record Society, 1995.

Cowley, Abraham. *A Proposition for the Advancement of Experimental Philosophy.* London: 1661.

Craig, John. *Newton at the Mint.* Cambridge: Cambridge University Press, 1946.

Crombie, A. C. *The History of Science from Augustine to Galileo.* Mineola, NY: Dover, 1995.

Crosby, Alfred W. *The Measure of Reality: Quantification and Western Society, 1250–1600*. Cambridge: Cambridge University Press, 1997.

Dalitz, Richard H.; and Nauenberg, Michael, eds. *The Foundations of Newtonian Scholarship*. Singapore: World Scientific, 2000.

David, F. N. 'Mr Newton, Mr Pepys & Dyse: A Historical Note'. *Annals of Science* 13 (1957): 137–47.

Davis, Philip J., and Hersh, Reuben. *The Mathematical Experience*. Boston: Houghton Mifflin, 1981.

De Morgan, Augustus. *Newton: His Friend: and His Niece*. Reprinted with introduction by E. A. Osborne. London: Dawsons, 1968.

de Santillana, Giorgio. *The Origins of Scientific Thought*. Chicago: University of Chicago Press, 1961.

de Villamil, Richard. *Newton the Man*. London: Gordon D. Knox, 1931.

Defoe, Daniel. *A Journal of the Plague Year*. Edited by Louis Landa. Oxford: Oxford University Press, 1998.

Dehaene, Stanislaus. *The Number Sense: How the Mind Creates Mathematics*. Oxford: Oxford University Press, 1997.

Descartes, René. *Philosophical Writings*. Translated by John Cottingham, Robert Stoothof and Dugald Murdoch. Cambridge: Cambridge University Press, 1984–91.

Ditchburn, R. W. 'Newton's Illness of 1692–3'. *Notes and Records of the Royal Society* 35 (1980): 1–16.

Dobbs, Betty Jo Teeter. *The Foundations of Newton's Alchemy, or 'The Hunting of the Greene Lyon'*. Cambridge: Cambridge University Press, 1975.

Dobbs, Betty Jo Teeter; and Jacob, Margaret C. *Newton and the Culture of Newtonianism*. Atlantic Highlands, NJ: Humanities Press, 1995.

Drake, Stillman. *Galileo at Work: His Scientific Biography*. Chicago: University of Chicago Press, 1978.

Dreyer, J. L. E. *A History of Astronomy from Thales to Kepler*. New York: Dover, 1953.

Eamon, William. *Science and the Secrets of Nature: Books of Secrets in Medieval and Early Modern Culture*. Princeton: Princeton University Press, 1994.

Easlea, Brian. *Witch Hunting, Magic and the New Philosophy: An Introduction to the Debates of the Scientific Revolution 1450–1750*. Sussex: Harvester Press, 1980.

Einstein, Albert. *Ideas and Opinions*. New York: Modern Library, 1994.

——. *Out of My Later Years*. New York: Carol, 1995.

Einstein, Albert; and Infeld, Leopold. *The Evolution of Physics*. New York: Simon & Schuster, 1938.

Eisenstein, Elizabeth L. *The Printing Press as an Agent of Change: Communications and Cultural Transformations in Early-Modern Europe*. Cambridge: Cambridge University Press, 1979.

Elliott, Paul. 'The Birth of Public Science in the English Provinces', *Annals of Science* 57 (2000): 61–100.

Elliott, Ralph W. V. 'Isaac Newton as Phonetician'. *Modern Language Review* 49 (1954): 1.

——. 'Isaac Newton's "Of an Universall Language"'. *Modern Language Review* 52 (1957): 1.

'Espinasse, Margaret. *Robert Hooke*. Berkeley: University of California Press, 1962.

Evelyn, John. *The Diary of John Evelyn*. Edited by E. S. de Beer. Six volumes. Oxford: Clarendon Press, 1955.

Fara, Patricia. *Newton: The Making of a Genius*. London: Macmillan, 2002.

Fauvel, John; and Gray, Jeremy, eds. *The History of Mathematics: A Reader*. London: Macmillan, 1987.

Fauvel, John; Flood, Raymond; Shortland, Michael; and Wilson, Robin, eds. *Let Newton Be!* Oxford: Oxford University Press, 1988.

Feingold, Mordechai, ed. *Before Newton: The Life and Times of Isaac Barrow*. Cambridge: Cambridge University Press, 1990.

Feynman, Richard. *The Character of Physical Law*. Introduction by James Gleick. New York: Modern Library, 1994.

Foster, C. W. 'Sir Isaac Newton's Family'. Reports and Papers of the Architectural & Archeological Society of the County of Lincoln 39 (1928).

Galileo Galilei. *The Controversy on the Comets of 1618*. Translated by Stillman Drake and C. D. O'Malley. Philadelphia: University of Pennsylvania Press, 1960.

——. *Dialogue Concerning the Two Chief World Systems – Ptolemaic & Copernican*. Translated by Stillman Drake, foreword by Albert Einstein. Berkeley: University of California Press, 1967.

——. *Discoveries and Opinions of Galileo*. Translated by Stillman Drake. New York: Anchor Books, 1957.

Garber, Daniel; and Ayers, Michael. *The Cambridge History of Seventeenth-Century Philosophy*. Cambridge: Cambridge University Press, 1998.

Gaule, John. *Pys-mantia the Mag-Astromancer, or the Magicall-Astrologicall-Diviner, Posed, and Puzzled*. London: J. Kirton, 1652.

Gjertsen, Derek. *The Newton Handbook*. London: Routledge & Kegan Paul, 1986.

Glanvill, Joseph. *Scepsis Scientifica: or, Contest Ignorance, the way to Science*. London: E. Cotes, 1665.

Goethe, Johann Wolfgang von. *Theory of Colours*. Translated by Charles Lock Eastlake, introduction by Deane B. Judd. Cambridge, Mass.: MIT Press, 1970.

Gooding, David; Pinch, Trevor; and Schaffer, Simon, eds. *The Uses of Experiment: Studies in the Natural Sciences*. Cambridge: Cambridge University Press, 1989.

Goodstein, David L.; and Goodstein, Judith R. *Feynman's Lost Lecture*. New York: Norton, 1996.

Gordon, George. *Remarks upon the Newtonian Philosophy*. London: W. W., 1719.

Grant, Edward. *The Foundations of Modern Science in the Middle Ages.* Cambridge: Cambridge University Press, 1996.

———. *Planets, Stars, and Orbs: The Medieval Cosmos, 1200–1687.* Cambridge: Cambridge University Press, 1994.

Greenstreet, W. J., ed. *Isaac Newton 1642–1727: A Memorial Volume.* London: G. Bell & Sons, 1927.

Guerlac, Henry. *Newton on the Continent.* Ithaca, NY: Cornell University Press, 1981.

Guicciardini, Niccolò. *The Development of Newtonian Calculus in Britain, 1700–1800.* Cambridge: Cambridge University Press, 1989.

———. *Reading the Principia: The Debate on Newton's Mathematical Methods for Natural Philosophy from 1687 to 1736.* Cambridge: Cambridge University Press, 1999.

Hall, A. Rupert. *Ballistics in the Seventeenth Century: A Study in the Relations of Science and War with Reference Principally to England.* Cambridge: Cambridge University Press, 1952.

———. *Isaac Newton: Eighteenth-Century Perspectives.* Oxford: Oxford University Press, 1999.

———. *Newton, His Friends and His Foes.* Aldershot, UK: Variorum, 1993.

———. *Philosophers at War: The Quarrel between Newton and Leibniz.* Cambridge: Cambridge University Press, 1980.

———. *The Scientific Revolution 1500–1800.* Second edition. Boston: Beacon Press, 1962.

Hall, A. Rupert; and Hall, Marie Boas. 'Newton's Theory of Matter', *Isis* 51 (March 1960): 163.

Hall, A. Rupert; and Hall, Marie Boas, eds. *The Correspondence of Henry Oldenburg.* Madison: University of Wisconsin Press, 1965–73.

Hall, Marie Boas. *Henry Oldenburg: Shaping the Royal Society.* Oxford: Oxford University Press, 2002.

Halley, Edmond. *Correspondence and Papers of Edmond Halley.* Edited by E. F. MacPike. Oxford: Oxford University Press, 1937.

Harman, P. M.; and Shapiro, Alan E., eds. *The Investigation of Difficult Things: Essays on Newton and the History of the Exact Sciences in Honour of D. T. Whiteside*. Cambridge: Cambridge University Press, 1992.

Harrison, John. *The Library of Isaac Newton*. Cambridge: Cambridge University Press, 1978.

Heilbron, J. L. *Physics at the Royal Society during Newton's Presidency*. Los Angeles: William Andrews Clark Memorial Library, 1983.

Hill, Christopher. *Change and Continuity in Seventeenth Century England*. Cambridge, Mass.: Harvard University Press, 1975.

History of Science Society. *Sir Isaac Newton 1727–1927: A Bicentenary Evaluation of His Work*. Baltimore: Williams & Wilkins, 1927.

Hobbes, Thomas. *Leviathan*. Edited by Richard E. Flathman and David Johnston. New York: Norton, 1996.

Hooke, Robert. *An Attempt to Prove the Motion of the Earth from Observations*. London: Royal Society, 1674.

——. *Diary, 1672–1680*. Edited by Henry W. Robinson and Walter Adams. London: Taylor & Francis, 1935.

——. *Lectures and Collections: Cometa* and *Microscopium*. London: J. Martyn, 1678.

——. *Micrographia: or some Physiological Descriptions of Minute Bodies Made by Magnifying Glasses with Observations and Inquiries thereupon*. London: J. Martyn and J. Allestry, 1665.

Hoppit, Julian. *A Land of Liberty? England 1689–1727*. Oxford: Oxford University Press, 2000.

Houghton, Walter E., Jr. 'The History of Trades: Its Relation to Seventeenth Century Thought'. *Journal of the History of Ideas* 2-1 (1941): 33–60.

Hubbard, Elbert. *Newton: Little Journeys to Homes of Great Scientists*. East Aurora, NY: Roycrofters, 1905.

Hunter, Michael, ed. *Robert Boyle Reconsidered*. Cambridge: Cambridge University Press, 1994.

Hunter, Michael; and Schaffer, Simon, eds. *Robert Hooke: New Studies*. Woodbridge, UK: Boydell Press, 1989.

Hutchinson, Keith. 'What Happened to Occult Qualities in the Scientific Revolution?' *Isis* 73 (1982): 233–53.

Huxley, G. L. 'Two Newtonian Studies'. *Harvard Library Bulletin* 13 (winter 1969): 348–61.

Iliffe, Robert. 'Playing Philosophically: Isaac Newton and John Bate's Mysteries of Art and Nature'. *Intellectual News* 8 (Summer 2000): 70.

Jacob, Margaret. *The Newtonians and the English Revolution, 1689–1720*. Ithaca, NY: Cornell University Press, 1976.

Jardine, Lisa. *Ingenious Pursuits: Building the Scientific Revolution*. New York: Doubleday, 1999.

Johns, Adrian. *The Nature of the Book: Print and Knowledge in the Making*. Chicago: University of Chicago Press, 1998.

——. 'Miscellaneous Methods: Authors, Societies, and Journals in Early Modern England'. *British Journal for the History of Science* 33 (2000): 159.

Johnson, L. W.; and Wolbarsht, M. L. 'Mercury Poisoning: A Probable Cause of Isaac Newton's Physical and Mental Ills'. *Notes and Records of the Royal Society* 34 (1979): 1.

Kaplan, Robert. *The Nothing That Is*. Oxford: Oxford University Press, 1999.

Klawans, Harold L. *Newton's Madness*. New York: Harper & Row, 1990.

Kollerstrom, Nicholas. *Newton's Forgotten Lunar Theory*. Santa Fe: Green Lion Press, 2000.

Koyré, Alexandre. *From the Closed World to the Infinite Universe*. Baltimore: Johns Hopkins University Press, 1957.

——. *Newtonian Studies*. Chicago: University of Chicago Press, 1965.

Kuhn, Thomas S. *The Structure of Scientific Revolutions*. Second edition. Chicago: University of Chicago Press, 1970.

Leedham-Green, Elisabeth. *A Concise History of the University of Cambridge*. Cambridge: Cambridge University Press, 1996.

Lenoir, Timothy, ed. *Inscribing Science: Scientific Texts and the Materiality of Communication*. Stanford, Calif.: Stanford University Press, 1998.

Li, Ming-Hsun. *The Great Recoinage of 1696 to 1699*. London: Weidenfeld & Nicolson, 1963.

Lindberg, David C. *The Beginnings of Western Science*. Chicago: University of Chicago Press, 1992.

Lindberg, David C.; and Westman, R. S., eds. *Reappraisals of the Scientific Revolution*. Cambridge: Cambridge University Press, 1990.

Lohne, Johannes A. 'Hooke versus Newton: An Analysis of the Documents in the Case on Free Fall and Planetary Motion'. *Centaurus* 7-1 (1960): 6–52.

——. 'Isaac Newton: The Rise of a Scientist', *Notes and Records of the Royal Society of London* 20-2: 125–39.

Lyons, Henry. *The Royal Society 1660–1940*. Cambridge: Cambridge University Press, 1944.

Mahoney, Michael S. 'The Beginnings of Algebraic Thought in the Seventeenth Century'. In S. Gaukroger, ed., *Descartes: Philosophy, Mathematics and Physics*. Sussex: Harvester, 1980.

Mancosu, Paolo. *Philosophy of Mathematics and Mathematical Practice in the Seventeenth Century*. Oxford: Oxford University Press, 1996.

Mandelbrote, Scott. *Footprints of the Lion: Isaac Newton at Work*. Cambridge: Cambridge University Library, 2001.

Manuel, Frank. *Isaac Newton, Historian*. Cambridge, Mass.: Harvard University Press, 1963.

——. *A Portrait of Isaac Newton*. Cambridge, Mass.: Harvard University Press, 1968.

McGuire, J. E. *Tradition and Innovation: Newton's Metaphysics of Nature*. Dordrecht: Kluwer, 1995.

McGuire, J. E.; and Rattansi, Piyo M., 'Newton and the "Pipes of Pan"'. *Notes and Records of the Royal Society* 21-2: 108–42.

McKnight, Stephen A., ed. *Science, Pseudo-Science, and Utopianism in Early Modern Thought*. Columbia: University of Missouri Press, 1992.

McLachlan, H., ed. *Sir Isaac Newton's Theological Manuscripts*. Liverpool: Liverpool University Press, 1950.

McMullin, Ernan. *Newton on Matter and Activity*. Notre Dame, Ind.: University of Notre Dame Press, 1978.

Meli, Domenico Bertoloni. *Equivalence and Priority: Newton versus Leibniz*. Oxford: Clarendon Press, 1993.

Merton, Robert K. *On the Shoulders of Giants: A Shandean Postscript*. The Post-Italianate Edition. Foreword by Umberto Eco. Afterword by Denis Donoghue. Chicago: University of Chicago Press, 1993.

——. 'Priorities in Scientific Discovery: A Chapter in the Sociology of Science'. *American Sociological Review* 22 (December 1957): 635–59.

——. *Science, Technology, & Society in Seventeenth Century England*. New York: Howard Fertig, 1970.

Moore, Jonas. *Moore's Arithmetick: Discovering the Secrets of that Art, in Numbers and Species*. London: Thomas Harper, 1650.

More, Henry. *An Antidote against Atheisme, or An Appeal to the Natural Faculties of the Minde of Man, whether there be not a God*. London: Roger Daniel, 1653.

More, Louis Trenchard. *Isaac Newton: A Biography*. New York: Scribner, 1934.

Moretti, Tomaso. *A General Treatise of Artillery: or, Great Ordnance*. Translated by Jonas Moore. London: Obadiah Blagrave, 1683.

Murdin, Lesley. *Under Newton's Shadow: Astronomical Practices in the Seventeenth Century*. Bristol: Adam Hilger, 1985.

Neugebauer, Otto. *The Exact Sciences in Antiquity*. New York: Dover, 1969.

Nicolson, Marjorie Hope. *Newton Demands the Muse: Newton's Opticks and the Eighteenth Century Poets*. Princeton: Princeton University Press, 1946.

——. *Science and Imagination*. Ithaca, NY: Great Seal Books, 1956.

Ornstein, Martha. *The Rôle of Scientific Societies in the Seventeenth Century*. Chicago: University of Chicago Press, 1928.

Palter, Robert, ed. *The Annus Mirabilis of Sir Isaac Newton: 1666–1966*. Cambridge, Mass.: MIT Press, 1970.

Park, David. *The Fire within the Eye*. Princeton: Princeton University Press, 1997.

Pepys, Samuel. *The Diary of Samuel Pepys*. Notes by Richard Lord Braybrooke. London: Dent, 1906.

Petty, William. *Political Arithmetick*. London: Robert Clavel, 1690.

Porter, Roy; and Teich, Mikuláš, eds. *The Scientific Revolution in National Context*. Cambridge: Cambridge University Press, 1992.

Price, Derek J. de Solla. 'Newton in a Church Tower: The Discovery of an Unknown Book by Isaac Newton'. *Yale University Library Gazette* 34 (1960): 124.

Pyenson, Lewis; and Sheets-Pyenson, Susan. *Servants of Nature: A History of Scientific Institutions, Enterprises, and Sensibilities*. New York: Norton, 1999.

Raphson, Joseph. *The History of Fluxions*. London: William Pearson, 1715.

Rattansi, Piyo M. *Isaac Newton and Gravity*. London: Wildwood, 1974.

Ronan, Colin A. *Edmond Halley: Genius in Eclipse*. Garden City, NY: Doubleday, 1969.

Royal Society. *Newton Tercentenary Celebrations*. Cambridge: Cambridge University Press, 1947.

Russell, Bertrand. *Mysticism and Logic*. New York: Norton, 1929.

Sabra, A. I. *Theories of Light from Descartes to Newton.* Cambridge: Cambridge University Press, 1981.

Schaffer, Simon. 'Newtonianism'. In Olby, R. C.; Cantor, G. N.; Christie, J. R. R.; and Hodge, M. J. S., eds., *Companion to the History of Modern Science.* London: Routledge, 1990.

——. 'Newton's Comets and the Transformation of Astrology'. In Patrick Curry, ed., *Astrology, Science and Society: Historical Essays.* Woodbridge, UK: Boydell Press, 1987.

Secord, James A. 'Newton in the Nursery: Tom Telescope and the Philosophy of Tops and Balls'. *History of Science* 23 (1985): 127.

Shapin, Steven. *The Scientific Revolution.* Chicago: University of Chicago Press, 1996.

——. *A Social History of Truth: Civility and Science in Seventeenth-Century England.* Chicago: University of Chicago Press, 1994.

Shapiro, Alan E. *Fits, Passions, and Paroxysms: Physics, Method, and Chemistry and Newton's Theories of Coloured Bodies and Fits of Easy Reflection.* Cambridge: Cambridge University Press, 1993.

——. 'The Gradual Acceptance of Newton's Theory of Light and Color, 1672–1727'. *Perspectives on Science* 4 (1996): 59–140.

Socolow, Elizabeth Anne. *Laughing at Gravity: Conversations with Isaac Newton.* Introduction by Marie Ponsot. Boston: Beacon Press, 1988.

Spargo, P. E.; and Pounds, C. A. 'Newton's "Derangement of the Intellect": New Light on an Old Problem'. *Notes and Records of the Royal Society* 34 (1979): 11–32.

Sprat, Thomas. *The History of the Royal Society.* Edited by Jackson I. Cope and Harold Whitmore Jones. London: Routledge & Kegan Paul, 1959.

Stayer, Marcia Sweet, ed. *Newton's Dream.* Kingston, Ont.: McGill-Queen's University Press, 1988.

Stewart, Larry. 'Other Centres of Calculation, or, Where the Royal

Society Didn't Count: Commerce, Coffee-houses and Natural Philosophy in Early Modern London'. *British Journal for the History of Science* 32 (1999): 133–53.

———. *The Rise of Public Science: Rhetoric, Technology, and Natural Philosophy in Newtonian Britain, 1660–1750.* Cambridge: Cambridge University Press, 1992.

Stimson, Dorothy. *Scientists and Amateurs: A History of the Royal Society.* London: Sigma, 1949.

Stuewer, Roger H. 'Was Newton's "Wave-Particle Duality" Consistent with Newton's Observations?' *Isis* 60 (autumn 1969): 203, 392–94.

Stukeley, William. *Memoirs of Sir Isaac Newton's Life, 1752.* Edited by A. Hastings White. London: Taylor & Francis, 1936.

Telescope, Tom (pseudonym). *The Newtonian System of Philosophy: Explained by Familiar Objects, in an Entertaining Manner, for the Use of Young Persons.* London: William Magnet, 1798.

Theerman, Paul; and Seeff, Adele F., eds. *Action and Reaction: Proceedings of a Symposium to Commemorate the Tercentenary of Newton's Principia.* Newark: University of Delaware Press, 1993.

Thomas, Keith. 'Numeracy in Early Modern England'. *Transactions of the Royal Historical Society* 37, 5th series (1987): 103–32.

Thorndike, Lynn. *A History of Magic and Experimental Science.* New York: Columbia University Press, 1923.

van Leeuwen, Henry G. *The Problem of Certainty in English Thought 1630–1690.* The Hague: Martinus Nijhoff, 1963.

Voltaire (François-Marie Arouet). *Letters on England.* Translated by Leonard Tancock. Harmondsworth, UK: Penguin, 1980.

Waller, Maureen. *1700: Scenes from London Life.* New York: Four Walls Eight Windows, 2000.

Wallis, John. *A Defence of the Royal Society, and the Philosophical Transactions, in Answer to the Cavils of Dr. William Holder.* London: Thomas More, 1678.

Walters, Alice N. 'Ephemeral Events: English Broadsides of Early Eighteenth-Century Solar Eclipses'. *History of Science* 37 (March 1999): 1–43.

Webster, Charles. *From Paracelsus to Newton: Magic and the Making of Modern Science.* Cambridge: Cambridge University Press, 1982.

Weinberg, Steven. *Facing Up.* Cambridge, Mass.: Harvard University Press, 2001.

Weld, Charles Richard. *A History of the Royal Society, with Memoirs of the Presidents.* London: 1848.

Westfall, Richard S. *Force in Newton's Physics: The Science of Dynamics in the Seventeenth Century.* London: Macdonald, 1971.

——. *Never at Rest: A Biography of Isaac Newton.* Cambridge: Cambridge University Press, 1980.

——. 'Newton and the Fudge Factor'. *Science* 179: 751.

——. *Science and Religion in Seventeenth-Century England.* New Haven: Yale University Press, 1958.

——. 'Short-Writing and the State of Newton's Conscience, 1662'. *Notes and Records of the Royal Society* 18 (1963): 10.

Whiston, William. *Memoirs of the Life and Writings of Mr. William Whiston.* Second edition. London: Whiston & White, 1753.

White, Lynn, Jr. *Medieval Technology and Social Change.* Oxford: Oxford University Press, 1962.

White, Michael. *Isaac Newton: The Last Sorcerer.* New York: Perseus, 1998.

Whiteside, D. T. 'The Expanding World of Newtonian Research'. *History of Science* 1 (1962): 16–29.

——. 'Isaac Newton: Birth of a Mathematician'. *Notes and Records of the Royal Society* 19 (1964): 53–62.

——. 'Newton's Early Thoughts on Planetary Motion: A Fresh Look'. *British Journal for the History of Science* 2 (December 1964): 117–37.

8

———. 'Newton's Marvellous Year: 1666 and All That'. *Notes and Records of the Royal Society* 21 (1966): 32.

Wilkins, John. *Mathematical and Philosophical Works.* Facsimile of 1708 edition. London: Frank Cass, 1970.

Yeo, Richard. *Encyclopaedic Visions: Scientific Dictionaries and Enlightenment Culture.* Cambridge: Cambridge University Press, 2001.

———. 'Genius, Method, and Morality: Images of Newton in Britain, 1760–1860'. *Science in Context* 2 (autumn 1988): 257.

Index

Page numbers from 199 onwards refer to notes; those in italics indicate captions.

269

P.S.

Ideas,
interviews
& features ...

Profile
Natasha Loder talks to James Gleick

IN 1993, JAMES GLEICK swore he would never write another biography. At that time, he had just completed an account of the exuberant life of Nobel Prize-winning physicist Richard Feynman – a project that took five years to complete. A decade after this promise, though, and Gleick has completed another biography. This time, however, the focus of his attention lies further back in history, with the life of mathematician and philosopher Isaac Newton.

When I asked Gleick what had attracted him to Newton's life, he replied that it was the surprising sense that the world Newton entered was 'darker and slower and more governed by ignorance than I had imagined'. For Gleick this realization opened the door to finding a new way of tackling Newton's life. 'I was gobsmacked to discover that Newton was the first man in his family to know how to read or write. And there he is, flipping back and forth between Latin and English and inventing the modern world. How is it possible for a person to make these jumps? And what does it say about the state of human culture or knowledge that a person born into a family of illiterate farmers can become Sir Isaac Newton?' The answer to this question is as hard to find as the answer to what makes genius. From Gleick's two biographies, the brilliant characters that emerge are obsessives and eccentrics with a peculiar way of looking at the world.

Ever since his first book – the now classic *Chaos: Making a New Science* in 1987 – Gleick has perhaps been best known as a science

writer. *Chaos* brought a new field of inquiry to the attention of the world, and launched Gleick's career as an author. So how do his two biographies fit in with his other work on scientific topics? It appears that the common thread, and perhaps secret, to his books is that they are all about people. He likes writing about people, he says. When he was working on *Chaos*, he continues, he couldn't separate the process of writing about the development of certain scientific ideas from the individual stories of the people who developed them. 'I felt I could tell a scientific story in an honest way through the people who were there.' Gleick's biography of Newton mostly tries to do the reverse, to tell the story of a person through his science. But, at the same time, Gleick cannot quite resist telling a scientific story in the background. This is a story about the nature of information in the seventeenth century, and how knowledge was transmitted and used to gain power and prestige.

Newton, says Gleick, was particularly difficult to write about because he died so very long ago. Although millions of words of Newton's writing were left behind, he has always been an elusive and shadowy character. But in this short book Gleick has extracted the telling details of this scientist's life. And the relative brevity of the biography makes it easier to see not merely the shadowy figure of Newton amongst a mass of information, but a concise picture painted of a man. In the process of writing this book, Gleick has obviously immersed himself in ▶

❛ I felt I could tell a scientific story in an honest way through the people who were there. ❜

Profile *(continued)*

◀ the details of Newton's life and work. I ask
him whether he enjoyed getting lost in
another person's life. 'Maybe,' he replies
elusively.

Gleick was born in New York City in
1954, and majored in English at Harvard –
graduating in 1976. He then went on to
found *Metropolis*, an alternative weekly
newspaper in Minneapolis, with two friends.
It was really fun, he says, an illustrious yet
short-lived venture. After the demise of
Metropolis, Gleick moved back to New York
and tried to get a job at the *New York Times*.
He succeeded. At first he worked on the
Metro desk, and later wrote features for the
Sunday magazine. His way on to the
magazine was through his interest in science.
As many journalists have discovered, it is
easier to get a foot in the door writing about
science – which many see as difficult – than
writing about most other subjects. Gleick's
trick, of course, was that he was interested in
the people behind the science. His first piece
for the Sunday magazine was a profile of
scientist Doug Hofstadter, who had had a
surprise bestseller with his book *Gödel,
Escher, Bach: an Eternal Golden Braid*.

In 1986, Gleick took leave of absence
from the *New York Times* for four months to
start work on his first book, *Chaos*. He
returned to the *Times* as a science reporter,
where he wrote for several years on
mathematics and physics – both notoriously
difficult areas to cover. Over the next year, he
crammed as much travel and reporting in as
he could, writing *Chaos* in the evenings and
weekends. In 1988, after the death of Richard

Feynman, Gleick left the *Times* to work on his biography. Since then, he has thrown off the chains of a regular desk job in journalism and is now an author and freelance writer. For some years, he wrote a freelance technology column for the *Times* magazine, and has also contributed to the *New Yorker* and the *Atlantic Monthly*. In 1993, he took the time to become, briefly, an Internet pioneer by founding The Pipeline, a New York City Internet service. His other books include *Faster: The Acceleration of Just About Everything*, and *What Just Happened: A Chronicle from the Information Frontier*.

Although Gleick has lived for most of his life among the hustle and bustle of New York City, today he lives in the rural calm of New York's scenic Hudson Valley. There in a house perched high above the river he lives with his wife, the journalist and author Cynthia Crossen, and their dog, Astro. ∎

❛ As many journalists have discovered, it is easier to get a foot in the door writing about science – which many see as difficult – than writing about most other subjects. ❜

A Critical Eye

IT IS A BRAVE biographer who attempts to weigh up the vast mass of Newton's achievements. Reviewers were generally in awe of James Gleick's deft and masterly handling of a subject considered by the **New Humanist** to be 'most unpromising'. Apparently, 'the standard view is that too little is known of Newton's life to make his genius understandable', but Gleick is commendable in his journalistic search for the human angle. 'He has read every document in sight, not only for the guts of what they say but for their incidental information.'

The **Financial Times** agrees that 'even the best biographers of Newton flirt with failure because of the utter remoteness of the subject. Since there are so few personal papers, his life is often conjecture.' But Gleick does not pursue the truth as voluminously as his subject, rather he 'keeps things simple … He is an elegant writer, brisk without being shallow, excellent on the essence of the work and revealing in his account of Newton's dealings with the times and with the men – admiring, dubious and downright hostile – with whom he condescended to interact.'

For the **Observer**, 'there are two inescapable conclusions to be drawn from Isaac Newton's life: that he was a prodigious genius of unsurpassed talent and a crazed ingrate of venomous self-obsession.' Revelling in the details of Newton's manically antisocial intellect and clandestine obsessions with alchemy and heretical theology, the **Observer** praises Gleick's concise and striking narrative as 'a perfect antidote to the many vast, bloated scientific biographies that currently flood the market'.

But the **Daily Telegraph** warns readers that Gleick is a tyrannical recruiter for the Newtonian cause. 'Anyone caught shying away from Newton is dismissed as another Romantic poet-type and a heretic to boot. And woe betide any historian of science who dares to suggest that in quantum theory or relativity there is anything that is not already anticipated or suspected or contained in the omniscient Newton.' Perhaps there is a touch of the divine in the scope of Newton's vision. As it is matter-of-factly put in the **Guardian**, 'it is not too much to say that our world was founded by Newton'. ∎

❛Perhaps there is a touch of the divine in the scope of Newton's vision.❜

Winner Takes All

By Natasha Loder

IT IS OFTEN said that history is written by the winners. Is the same true of science? While history often hinges on bloody battles of conquest, some might argue that science is above all this – being only the high-minded pursuit of truth. But science, too, can be a battle: of priority and of competing visions. Although scientific ideas are rarely born intact, and rely on the work of others, not everyone who makes an important contribution emerges from the scientific trenches to be remembered by history. Scientific truth may win out in the end, but the spoils are not always distributed fairly.

Isaac Newton was certainly just such a scientific winner. From his conception of the law of universal gravitation, to his discoveries in mathematics, optics and astronomy, he was one of the finest minds of his age. Is it possible, though, that this extraordinary figure eclipsed many other fine minds of the time?

Newton is buried in Westminster Abbey in London, only a few feet away from another scientific luminary, Charles Darwin. Darwin is remembered as the father of evolutionary biology. But there was another man who can rightfully be called the co-discoverer of evolution by natural selection. His name is unfamiliar. And while Darwin rests at a grand London address, Alfred Russel Wallace can be found in a remote grave in Broadstone, Dorset. Five years ago, a group was set up to raise funds to restore his neglected and crumbling grave.

As James Gleick's book records, Newton also managed to eclipse the work of several others. Robert Hooke was one of them. Some believe he lost credit for an important idea because he argued with Newton, and lost. While it is difficult to say to what extent Newton actively played a role in Hooke's erasure from history, there is certainly something odd about the way that history has forgotten him.

Hooke was a brilliant experimental scientist and designer. In 1665 he was also a well-known figure in general society, largely because of his bestselling book *Micrographia*. This contained accurate and stunning drawings of what he saw under the microscope. For the first time people were able to see a world that was previously invisible to the naked eye. The structure of a hair, and the patterns made by a snowflake, were extraordinary revelations at the time.

Today Hooke is the subject of an almost apologetic revival based on his relative obscurity. The Royal Society, Britain's leading scientific academy, recently hosted a conference entitled 'Restoring the Reputation of Robert Hooke'. And a biography, *The Curious Life of Robert Hooke: The Man Who Measured London*, by Lisa Jardine, argues that he played an equal role in many of the projects attributed to Christopher Wren – in particular the dome of St Paul's Cathedral. ▶

❝ Newton also managed to eclipse the work of several others. Robert Hooke was one of them. ❞

Winner Takes All *(continued)*

◄ Although Newton started out his work as a scientific recluse in Cambridge, towards the middle and end of his life he became a powerful scientific figure and increasingly controlled a great deal of the institutional machinery within science. Hooke is not the only person whose reputation is less than might have been expected.

Rebekah Higgitt, of Imperial College in London, is a researcher interested in the significance of Newton's image. She says Newton's supporters helped him to write the history of science in a number of ways. Those whose reputations suffered included the astronomer John Flamsteed and Gottfried Leibniz, the philosopher and mathematician who is today known for having invented differential and integral calculus independently of Newton. Both the characters and scientific work of these men were criticized. By the nineteenth century, she adds, these stories were beginning to be modified. This was the result of new research and new perceptions of Newton's genius, but also because of 'resentment at the way in which Newton's reputation was allowed to sweep away all before it'.

Nevertheless, what historians write and what the public remembers can be different things. Scientists such as Newton and Darwin were, without question, outstanding scientists. But because we tend to paint our heroes in primary colours, some scientists will be remembered at the expense of others. And this is why Gleick's account of Newton is so valuable. It is a short and meticulous portrait of the man painted in every shade

❝Newton's supporters helped him to write the history of science in a number of ways. Those whose reputations suffered included the astronomer John Flamsteed and Gottfried Leibniz, ❞

possible. It is clear that Newton was a giant, a scientific mind beyond all others of the time. But he was also a real person, and so prone to all the peculiarities and vanities that one might expect.

Natasha Loder, Science and Technology Correspondent, *Economist* ∎

‘It is clear that Newton was a giant, a scientific mind beyond all others of the time. But he was also a real person, ’

Have You Read?

What Just Happened: A Chronicle from the Information Frontier
Previously published essays by Gleick which form an eclectic chronicle of the information revolution's first ten years.

....

Faster: The Acceleration of Just About Everything
Gleick dissects with acute insight and mordant wit our unceasing daily struggle to squeeze as much as we can – but never enough – into the 1,440 minutes of each day and shows the biological, psychological and neurological limits of just how much we are capable of doing.

....

Chaos: Making a New Science
This book brings together different work in the new field of physics called the chaos theory, an extension of classical mechanics, in which simple and complex causes are seen to interact. Mathematics may only be able to solve simple linear equations which experiment has pushed nature into obeying in a limited way, but now that computers can map the whole plane of solutions of non-linear equations a new vision of nature is revealed. The implications are staggeringly universal in all areas of scientific work and philosophical thought. Focuses on the personalities studying chaos as much as chaos itself.

Genius: The Life and Science of Richard Feynman

For nearly 50 years, until his death in 1988, Richard Feynman's work lay at the heart of the development of modern physics. Always controversial, Feynman was the key physicist of his time, from his work as part of the A-bomb-making team at Los Alamos in the early 1940s until his discovery of the reason for the Challenger space shuttle disaster 40 years later. The book combines biography with an accessible account of his thought and its context. ■

If You Loved This,
You Might Like ...

The Lunar Men: The Friends Who Made the Future
Jenny Uglow
A highly acclaimed portrait of the Lunar Society of Birmingham, the group of friends who launched the Industrial Revolution.

The Curious Life of Robert Hooke: The Man Who Measured London
Lisa Jardine
The biography of a brilliant, forgotten maverick. An engineer, surveyor, architect and inventor, Robert Hooke was a major figure in the seventeenth-century cultural and scientific revolutions and Isaac Newton's greatest rival.

$E=mc^2$: A Biography of the World's Most Famous Equation
David Bodanis
An energetic and accessible exploration of what came before and after the twentieth century's most revolutionary discovery.

Fermat's Last Theorem
Simon Singh
The remarkable story of one man's lifelong fascination with a seemingly simple challenge, Fermat's Last Theorem, a puzzle which has confounded mathematicians for centuries.

Newton: The Making of a Genius
Patricia Fara
Although Newton is now considered one of
the 'Great Britons', his genius has not always
been universally lauded. Fara assesses
Newton's posthumous reputation in the eyes
of the world's scientific, artistic and literary
communities.

*Mendeleyev's Dream: The Quest for
the Elements*
Paul Strathern
From the four elements of the Greeks to
Mendeleyev's discovery of the Periodic Table,
Strathern recounts the historical drama of
chemical science.

Never at Rest: A Biography of Isaac Newton
Richard S. Westfall
The authoritative scientific biography of
Newton, rich and painstaking in detail. ∎

Find Out More

www.around.com
James Gleick's home site.

http://www.newton.cam.ac.uk/newton.html
A brief survey of Newton's life and works and a useful list of links provided by the Isaac Newton Institute for Mathematical Sciences.

http://www-gap.dcs.st-and.ac.uk/~history/PictDisplay/Newton.html
A range of portraits of Newton past and present, including paintings, engravings, Newton's death mask, images on stamps and banknotes.

http://www.npg.org.uk/live/search/person.asp?search=ss&sText=Isaac+Newton&LinkID=mp03286
Portraits of Newton held by the National Portrait Gallery.

http://www.nationaltrust.org.uk/scripts/nthandbook.dll?ACTION=PROPERTY&PropertyId=82
Visiting information for National Trust-owned Woolsthorpe Manor, Newton's birthplace and family home.

http://www.newtonproject.ic.ac.uk/
The Newton Project – an electronic archive of Newton's manuscripts.

http://csep10.phys.utk.edu/astr161/lect/history/newtongrav.html
An accessible and entertaining explanation of the theory of universal gravitation.

http://www.phys.virginia.edu/classes/109N/
more_stuff/Applets/newt/newtmtn.html
Fire Newton's cannonball! Weaker shots fall
in parabola, but soon the curvature of the
earth becomes more important and stronger
shots orbit the earth in ellipses.

http://news.bbc.co.uk/1/hi/health/2988647.
stm

http://www.newscientist.com/news/news.
jsp?id=ns99993676
Two articles relating to the theory that both
Newton and Einstein suffered from
Asperger's Syndrome.

http://education.guardian.co.uk/print/
0,3858,4546885-108228,00.html
Tristram Hunt looks at the influence of
Newton's rational scientific method on
progressive ideals. ■

BOOKSHOP

Now you can buy any of these great paperbacks from Harper Perennial at **10%** off recommended retail price. *FREE postage and packaging in the UK.*

Fermat's Last Theorem
Simon Singh (ISBN: 1–84115–791–0)   £8.99
....

The Code Book
Simon Singh (ISBN: 1–85702–889–9)   £8.99
....

Longitude
Dava Sobel (ISBN: 1–85702–571–7)   £6.99
....

Galileo's Daughter
Dava Sobel (ISBN: 1–85702–712–4)   £7.99
....

The Queen's Conjuror: the Science and Magic of Dr Dee
Benjamin Woolley (ISBN: 0–00–655202–1)   £9.99
....

Nature via Nurture
Matt Ridley (ISBN: 1–84115–746–5)   £8.99
....

The Music of the Primes
Marcus du Sautoy (ISBN: 1–84115–580–2)   £8.99
....

The Curious Life of Robert Hooke
Lisa Jardine (ISBN: 0–00–715175–6)   £8.99
....

	Total cost
10% discount	
Final total	

*To purchase by Visa/Mastercard/Switch simply call **08707 871724** or fax on **08707 871725***

To pay by cheque, send a copy of this form with a cheque made payable to 'HarperCollins Publishers' to: Mail Order Dept (Ref: B0B4), HarperCollins Publishers, Westerhill Road, Bishopbriggs, G64 2QT, making sure to include your full name, postal address and phone number.

From time to time HarperCollins may wish to use your personal data to send you details of other HarperCollins publications and offers. If you wish to receive information on other HarperCollins publications and offers please tick this box ☐

Do not send cash or currency. Prices correct at time of press. Prices and availability are subject to change without notice. Delivery overseas and to Ireland incurs a £2 per book postage and packing charge.